Seoul Mini-Mountains

서울의 작은 산

Géraldine Borio 제랄딘 보리오

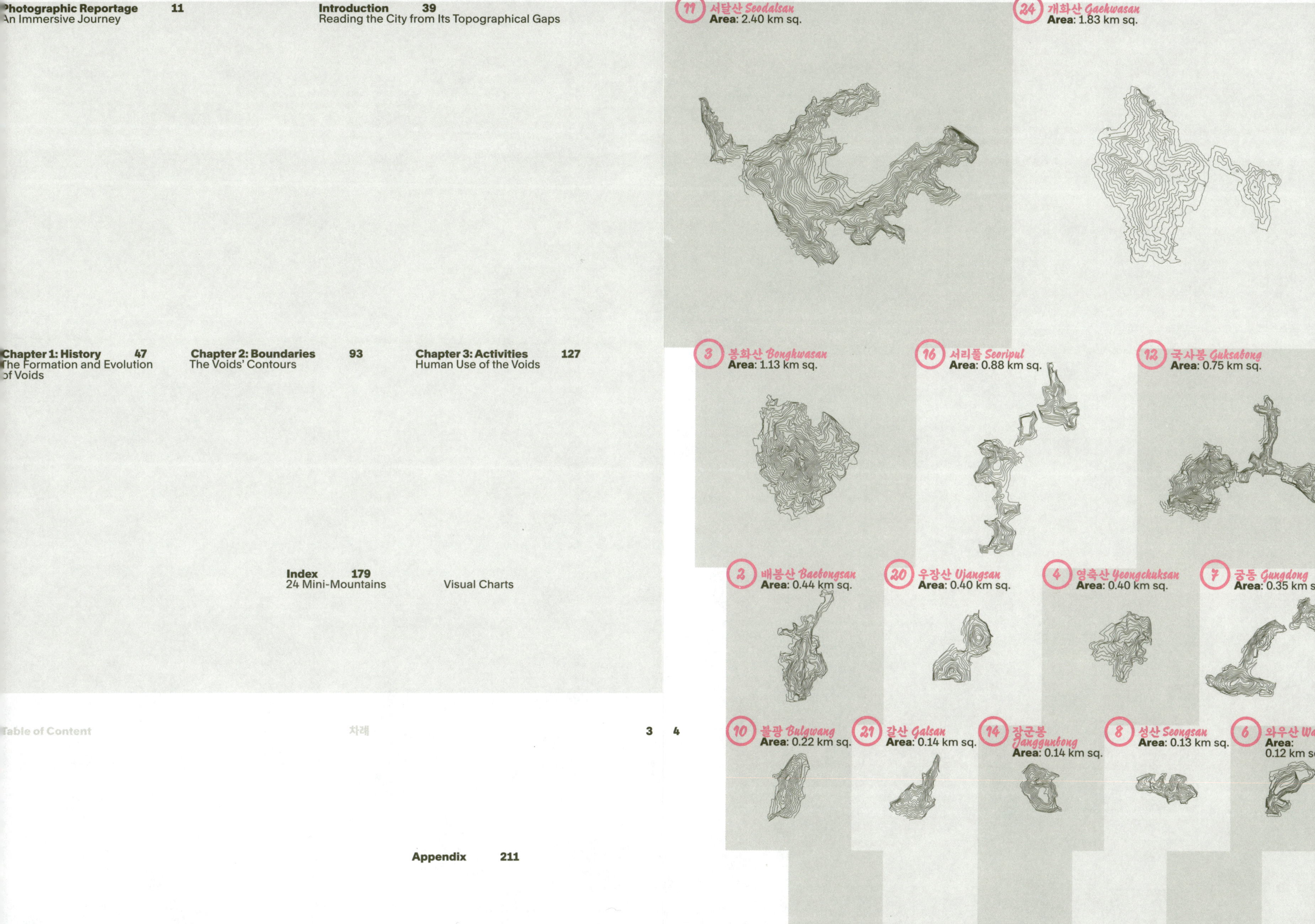

11 서달산 Seodalsan
Area: 2.40 km sq.
24 개화산 Gaehwasan
Area: 1.83 km sq.
3 봉화산 Bonghwasan
Area: 1.13 km sq.
16 서리풀 Seoripul
Area: 0.88 km sq.
12 국사봉 Guksabong
Area: 0.75 km sq.
2 배봉산 Baebongsan
Area: 0.44 km sq.
20 우장산 Ujangsan
Area: 0.40 km sq.
4 영축산 Yeongchuksan
Area: 0.40 km sq.
7 궁동 Gungdong
Area: 0.35 km sq.
10 불광 Bulgwang
Area: 0.22 km sq.
21 갈산 Galsan
Area: 0.14 km sq.
14 장군봉 Janggunbong
Area: 0.14 km sq.
8 성산 Seongsan
Area: 0.13 km sq.
6 와우산 Wau
Area: 0.12 km sq.

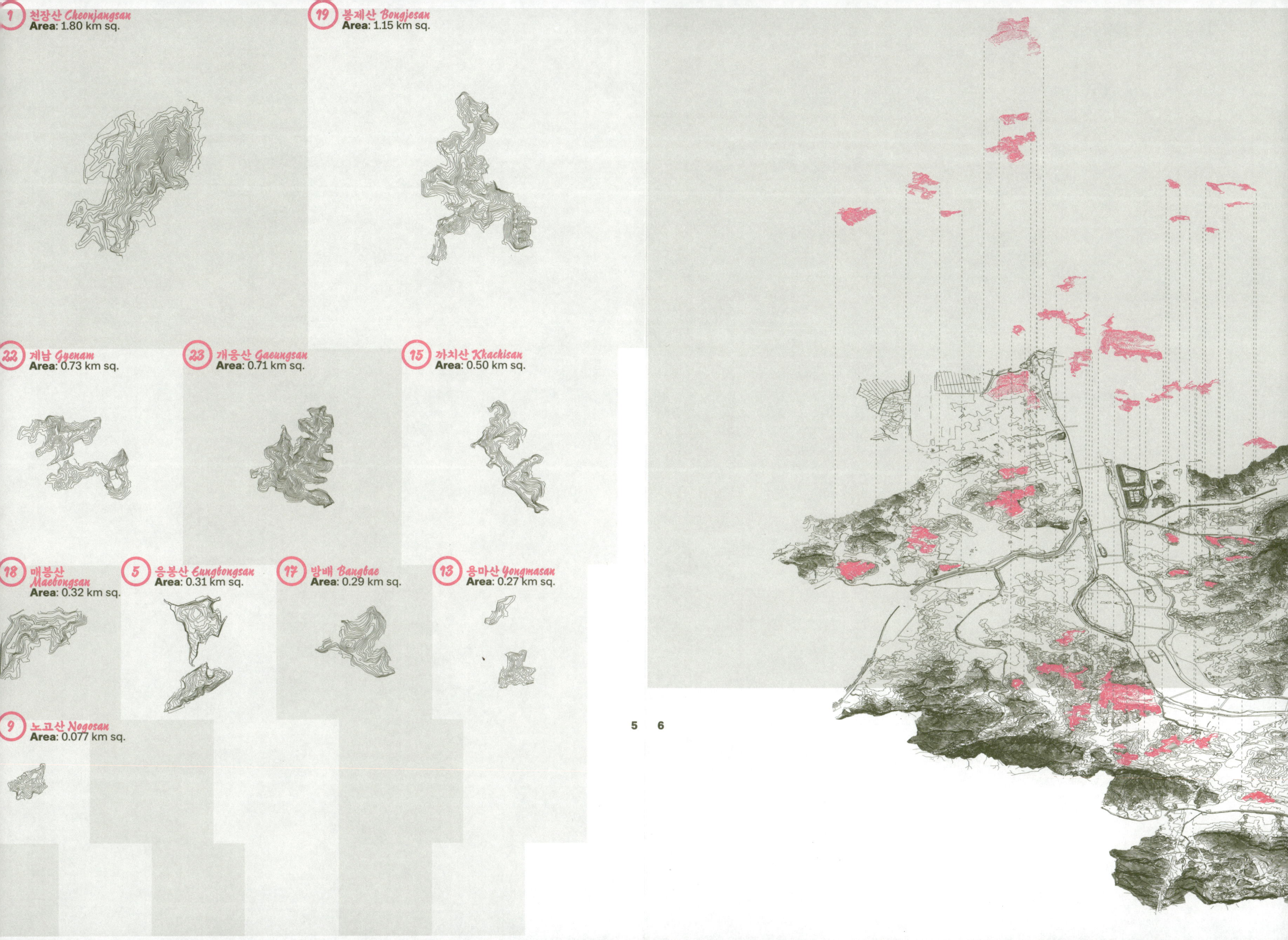

1 천장산 Cheonjangsan
Area: 1.80 km sq.
19 봉제산 Bongjesan
Area: 1.15 km sq.
22 계남 Gyenam
Area: 0.73 km sq.
23 개웅산 Gaeungsan
Area: 0.71 km sq.
15 까치산 Kkachisan
Area: 0.50 km sq.
18 매봉산 Maebongsan
Area: 0.32 km sq.
5 응봉산 Eungbongsan
Area: 0.31 km sq.
17 방배 Bangbae
Area: 0.29 km sq.
13 용마산 Yongmasan
Area: 0.27 km sq.
9 노고산 Nogosan
Area: 0.077 km sq.

1
Distribution of Mini-Mountains in Seoul 서울의 작은 산 분

There are 68 mini mountains located in the urban area in Seoul. 24 of them are chosen for further analysis in regard to their significance to the daily life of the inhabitants.

서울에는 총 68곳의 작은 산이 있다. 이 중 24곳은 주민의 일상생활에서 갖는 중요성을 고려하여 추가 분석 대상으로 선정했다.

1 천장산 Cheonjangsan
2 배봉산 Baebongsan
3 봉화산 Bonghwasan
4 영축산 Yeongchuksan
5 응봉산 Eungbongsan
6 와우산 Wausan
7 궁동 Gungdong
8 성산 Seongsan
9 노고산 Nogosan
10 불광 Bulgwang
11 서달산 Seodalsan
12 국사봉 Guksabong
13 용마산 Yongmasan
14 장군봉 Janggunbong
15 까치산 Kkachisan
16 서리풀 Seoripul
17 방배 Bangbae
18 매봉산 Maebongsan
19 봉제산 Bongjesan
20 우장산 Ujangsan
21 갈산 Galsan
22 계남 Gyenam
23 개웅산 Gaeungsan
24 개화산 Gaehwasan

사진 르포르타주

An Immersive Journey

구석구석 산책하기

I'PARK

95
대문앞

615
신내
614
613

130
127

101
금연구역

봉천로23바길
Bongcheon-ro 23ba-gil

진리가 너희를
자유롭게 하리라

09

족욕장

머리말

Reading the City from Its Topographical Gaps

지형 격차로 도시 읽기

『서울의 작은 산』은 도시에 남겨진 보이드(voids, 빈 공간)의 역할과 그것이 도시 내러티브와 정체성에 미치는 영향을 탐구함으로써 서울을 재조명하고자 한다. 서울의 독특한 지형 덕분에 개발되지 않은 작은 산 스물네 곳은 인구와 건축물이 밀집한 도시 구조를 방해하고 있다. 오늘날 서울 수도권을 위에서 바라보면 66–200미터 높이의 작은 돌출부들이 빽빽한 도시에 별자리처럼 흩어져 건물 사이사이에 보이드를 이루고 있다. 도시화의 흐름에 저항하는 이런 개발 패턴은 도시가 숨 쉴 수 있게 한다.

명확한 경계가 없는 평지 위의 도시에서 교외 지역은 보통 외곽에 있다. 그러나 서울은 특별한 지형 덕분에 일부 지역이 개발되지 않은 채 도심에 남아 있다. 이렇게 남겨진 완충 지대는 주민들의 일상에 깊숙이 자리 잡으며 빈틈없는 도시에서 벗어나 휴식할 수 있는 공간을 제공한다. 하지만 완충 지대는 중요한 랜드마크로 여겨지지 않기 때문에 도시 개발이 급격히 확장함에 따라 서서히 잠식되기 쉬우며, 끝내 소중한 완충 기능을 상실할 우려가 있다.

이 연구는 서울의 작은 산 스물네 곳을 중점적으로 조사함으로써 도시를 이해하고 탐구하는 독특한 시각을 제시한다. 이들 작은 산은 새로운 도시 내러티브를 형성하고 종종 간과되는 미개발 공간의 중요성을 부각한다. 도시를 단순히 건축물의 집합체로 보는 보편적 인식에 도전할 뿐 아니라 보이드의 가치를 강조한다. 또한 역사 속에서 형성되어 온 과정과 개발 경계(도시화된 지역과 미개발 지역을 구분하는 경계선)의 변화, 그리고 미개발 지역에서 일어나는 활동을 분석함으로써 보이드를 연구하고 서울이라는 도시의 진화 과정을 새로운 관점에서 살펴본다. 이렇게 남겨진 작은 산을 통해 도시를 새롭게 탐구하며 보이드의 군도로 둘러싸인 도시의 또 다른 모습을 드러내 보인다.

Seoul Mini-Mountains: Reading the City from Its Topographical Gaps aims to rethink the city of Seoul by examining the role of residual urban voids and their impact on urban narratives and identities. Thanks to the city's specific geography, two dozen undeveloped mini-mountains interrupt the dense city fabric. Nowadays, when looking at the metropolitan area of Seoul from above, small protrusions are apparent, ranging from 66 to 200 meters high. These appear like a constellation of urban voids framed by a mass of buildings. This pattern of development with small areas that seem to resist the tide of urbanisation is repeated across the city, allowing it to breathe.

In flat cities without clear territorial boundaries, the suburbs are usually found at the city's periphery. The specific geography of Seoul, however, has resulted in portions of undeveloped ground within the city itself. These residual buffer zones are embedded within daily life, often offering inhabitants a respite from dense urbanisation. Not considered major cultural landmarks, they are under threat of being gradually eaten up by the frenetic urban sprawl, with the concomitant risk of losing their valuable buffer qualities.

This investigation focuses on a selection of 24 mini-mountains in Seoul, each offering a distinct lens through which to read and understand the city. Constructed from these urban fragments, a new narrative of the city emerges—one that highlights the importance of unbuilt and often overlooked spaces. This study challenges the common perception of a city as merely an accumulation of built forms, instead emphasizing the value of these voids. By analysing these urban voids through their formation during the *History*, the transformation of their built *Boundaries*, and the observation of *Activities* within their unbuilt territories, the study provides original insights into the city's evolution. Interrogating a new reading of a city via its overlooked and residual mini-mountains now brings forward another image of the city: one that revolves around an archipelago of voids.

The core motivation of this investigation is the rethinking of what a city is, through the lens of specific urban conditions: buffer zones, in-between spaces and urban thresholds, all of them by-products of larger development agendas. Looking at Seoul through the lenses of small topographical buffers expands the notion of spatial ambiguity at the geographical scale, contributing to the construction of a specific type of urban knowledge.

이 연구는 완충 지대, 사이 공간, 도시 경계(도시의 한 영역에서 다른 영역으로 전환되는 지점) 같은 특정 조건을 통해 도시란 무엇인가를 재고하는 데 초점을 맞추었다. 이들 모두는 더 광범위한 개발계획의 부산물이다. 소규모 지형적 완충 지대를 통해 서울을 관찰하면 지리적 차원에서 도시 공간은 더욱 모호해지며, 이는 도시와 관련한 특정 유형의 지식을 형성하는 데 기여한다.

이러한 작업은 도시를 걷다가 서울의 전체적인 구조와는 어딘가 어울리지 않아 보이는 녹지 공간들을 발견하면서 시작되었다. 일반적으로 도시의 이미지는 상징적인 건물, 웅장한 광장, 눈에 띄는 경관 등 쉽게 식별할 수 있는 랜드마크를 중심으로 구성된다. 서울은 북악산, 인왕산, 남산, 낙산이라는 네 곳의 주요 산으로 구시가지 공간 구조의 본질을 이해할 수 있다. 도시는 개별적으로는 서로 무관해 보이는 작은 공간들이 우연히 축적되면서 형성된 결과물이기도 하다. 도시라는 체계 안에 존재하는 이런 단편적인 공간들을 살펴보면 다양한 변형 패턴이 드러나고, 도시를 또 다른 시각으로 이해할 수 있다. 역사학자 카를로 긴츠부르그의 말처럼, 표준은 결코 이상 현상을 예측할 수 없지만 이상 현상은 표준을 예측할 수 있다.

첫 번째로 서울의 작은 산 스물네 곳과 그 주변을 개발, 보이드, 지형, 접근성, 기능, 활동, 식생, 법적 구분 등 여러 층으로 나누어 분석했다. 각 작은 산을 구글 어스, GIS, 역사적 지도, 토지이용 및 규제를 보여주는 법적 지도 등을 사용해 세밀히 분석했고, 이렇게 얻은 여러 정보를 통합하여 다시 그린 후 체계적으로 분류하고 비교했다. 두 번째 단계에서는 현장을 방문해 지방자치단체와 주민들이 작은 산을 어떻게 사용하고 변형하는지 자세히 관찰했다. 이 두 가지 접근 방식으로 수집한 자료를 결합해 작은 산의 형성과 변화 과정을 한눈에 파

This work started with walking in the city and noticing pockets of green space that didn't seem to be in keeping with Seoul's overall layout. The general image of a city is often constructed around identifiable landmarks: iconic buildings, grandiose plazas, remarkable landscape features. In the case of Seoul, the four main inner-Seoul urban mountains Bugaksan 북악산, Inwangsan 인왕산, Namsan 남산, and Naksan 낙산 were an entry point to the understanding of the underlying organisation of the old city. But a city is also an accumulation of smaller spatial accidents that seem irrelevant when looked at individually. When examining such fragments in a system, revealing their different transformation patterns, an alternative reading of the city becomes possible. As the historian Carlo Ginzburg puts it: *The norm can never predict the anomaly, however, the anomaly can predict the norm.*

During the first phase of investigations, the 24 sites and their surroundings were decomposed according to a series of layers (built, void, topography, accessibility, function, activities, vegetation, legal division). Each mini-mountain was dissected using Google Earth, GIS, historical and legal maps, and then redrawn, incorporating multiple layers of information, allowing a systematic classification and comparison. In a second phase, a series of detailed on-site, immersive observations focussed on the ways in which sites are used and modified by their users and the districts' governments. These two approaches led to the re-drawing of those sites, combining off-site and on-site data, to get a panoptical understanding of the mechanism of formation and transformation of these spaces.

The first discovery was that the mini-mountains were not isolated events but a phenomenon. The analysis reveals the interplay between contrasting forces: the social and economic transformations of the country and top-down urban planning strategies. For example, while many mini-mountains appear natural today, several have undergone significant occupation during peak urbanisation periods, such as mountains 5 응봉산 Eungbongsan, and 12 국사봉 Guksabong. Some were incorporated early into the city's development agendas for green spaces, 7 궁동 Gungdong; others were left on the side but remaining as valuable buffers for the neighbourhood inhabitants. Some, like in 20 우장산 Ujangsan, are deeply

악하도록 다시 그릴 수 있었다.

가장 처음 발견한 사실은 서울의 작은 산 스물네 개는 각각 개별적인
사건이 아니라 전체가 하나의 현상을 이룬다는 점이었다. 한국의 사회적, 경제적
변화와 하향식 도시계획 전략 사이에서 발생하는 상반된 힘들이 상호작용한다는
분석 결과가 드러났다. 예를 들면 대부분의 작은 산이 원래부터 지금 모습이었던
것처럼 보이지만 응봉산 5 과 국사봉 12 등 몇몇 산은 도시화가 절정에
달했을 때 많은 주민의 거주지였다. 궁동 7 처럼 도시 개발계획에 녹지 공간
으로 포함되었던 산도 있고, 방치되었지만 지역 주민에게 중요한 완충 지대로
기능했던 산도 있다. 우장산 20 같은 일부 작은 산은 지역 주민의 일상에 깊
이 뿌리내려 다양한 야외 활동을 가능케 하고 자연으로의 탈출구를 제공한다.
천장산 1 같이 높은 건물에 둘러싸여 접근하기 어려운 산은 주민이 일상적
으로 사용할 수 없어 도심 속 '비밀 정원'처럼 기능한다. 자세히 들여다보면 몇몇
작은 산은 도시 주민에게 서울이라는 도시와 자신이 어떤 관계를 맺고 있는지
느끼게 해준다. 예컨대 산에 깃든 정기와 함께 개화산 24 은 서울의 역사에서
오랫동안 함께해 왔다. 또 최근 맨발 걷기 같은 체험이 유행하는 데서 알 수 있
듯, 작은 산은 도시 주민과 자연환경 사이를 이어주기도 한다. 조금 더 큰 틀에서
보면 이런 공간은 도시 생태계에 꼭 필요하다. 이미 주어진 천연 자산으로, 도시
를 계획할 때 대규모 재개발이 필요 없는 귀중한 자원이다.

처음에 이 연구는 도시 속 보이드(빈 공간)를 살펴보는 데서 출발했
지만, 조사할수록 보이드라는 지형 요소는 사실 빈 공간이 아님을 알게 되었다
작은 산들은 솔리드(solid)와 보이드, 사적 공간과 공유 공간, 야생과 조성된 공
간 사이에서 전환이 지속적으로 이루어지는 역동적인 변화 체계를 제공한다. 이

connected to the daily lives within the neighbourhood, offering a range of outdoor activities and wild escapes. Others, due to surrounding tall buildings and lack of access, seem not to exist within the mental map of the residents, making them 'secret gardens' in the urban landscape, (1) 천장산 *Cheonjangsan*. At the micro level, some mini-mountains enable city inhabitants to connect with their identity—these mountains and their spirits have been integral to the city's history, like (24) 개화산 *Gaehwasan*—and with their environment, as seen in the recent trend of walking barefoot and feeling the ground. At the macro scale, these spaces are vital for the ecosystem of the city. They are ready-made natural assets and offer urban planners valuable resources that do not require massive redevelopment.

If the premise of this research initially appeared to be an exploration of voids within the city, the investigation revealed that these topographic features are far from empty. The findings suggest that the mini-mountains offer Seoul a dynamic system of transitional spaces, continually oscillating between solid and void, private and shared, wild and constructed. Each mini-mountain occupies an ambiguous space within this system, inviting varied interpretations of the physical and symbolic boundaries that define the city's landscapes.

Examining the tangible and intangible traces left by history, as well as the accumulation of multiple layers—programmatic, spiritual, political, legal, built, and unbuilt—demonstrated the depth of significance encapsulated within these territories. Seoul's mini-mountains deserve recognition as emblematic features of the city's contemporary identity—an identity that resides not only in the built environment but also unfolds atop these hills.

Though deeply woven into the city's fabric, the mini-mountains remain fragile. Even if we choose not to build upon them, let's resist the temptation to turn them into amusement parks. Their beauty lies in their partial escape from human control. Before making any interventions, we should observe closely and listen attentively; like living entities, these spaces will continue to evolve over time. This book does not propose an urban strategy, nor does it offer yet another development plan to dictate their status. Instead, it invites you to pause, to notice what makes this city truly unique—those intangible qualities that quietly shape its identity.

런 체계 속에서 작은 산들이 차지하는 공간은 모호한 성격을 띠며, 도시경관을 정의하는 물리적이고도 상징적인 경계를 다양하게 해석할 여지를 남긴다.

역사가 남긴 유무형의 흔적뿐 아니라 지금까지 축적되어 온 계획적, 영적, 정치적, 법적, 개발적 차원의 다양한 층위를 살펴보면 이런 공간에 담긴 심오한 의미가 드러난다. 서울의 작은 산은 단순히 지형의 일부를 넘어 오늘날 서울의 정체성을 상징하는 중요한 요소로 인정받아 마땅하다. 그 정체성은 건축물들 사이에만 있는 것이 아니라, 작은 산들의 언덕 위에서도 서서히 드러난다.

도시의 조직에 깊이 얽혀 있는 서울의 작은 산은 여전히 취약한 존재다. 산중에 개발 사업을 진행하지 않는 것은 물론이고, 놀이공원을 세우려는 유혹도 물리쳐야 한다. 이 산의 아름다움은 인간의 통제를 벗어난 영역의 일이기 때문이다. 자연에 개입하기 전에 우리는 먼저 자세히 관찰하고 귀 기울여 들어야 한다. 작은 산들은 살아 있는 존재처럼 시간이 흐르면서 자연히 진화해 나갈 것이다. 이 책은 도시 전략을 제시하지 않으며, 또 다른 개발계획을 통해 이들 산의 지위를 정하려 하지 않는다. 그저 잠시 멈춰 서서, 서울을 진정으로 특별한 도시로 만드는 것, 즉 도시의 정체성을 묵묵히 형성하는 보이지 않는 것들을 떠올려 보기를 권한다.

Chapter 1

1 장

History
역사

The Formation and Evolution of Voids
보이드의 형성과 변화

오늘날 도시에서 보이드가 형성되고 변화해 온 과정을 이해하려면 역사 기록물을 바탕으로 도시를 형성한 유무형의 요소를 함께 살펴보아야 한다. 오래된 지형도를 참조하면 작은 산의 역사를 파악하고, 아직 개발되지 않은(남겨진) 공간이 어떤 층위를 거쳐 형성되었는지 볼 수 있다. 그리고 이런 모습은 상반된 힘의 역동적인 상호작용을 드러낸다. 어쩌면 도시화의 물결과 이를 막아내려는 산의 숨결, 혹은 산의 정기가 끊임없이 밀고 당기는 것은 아닐까?

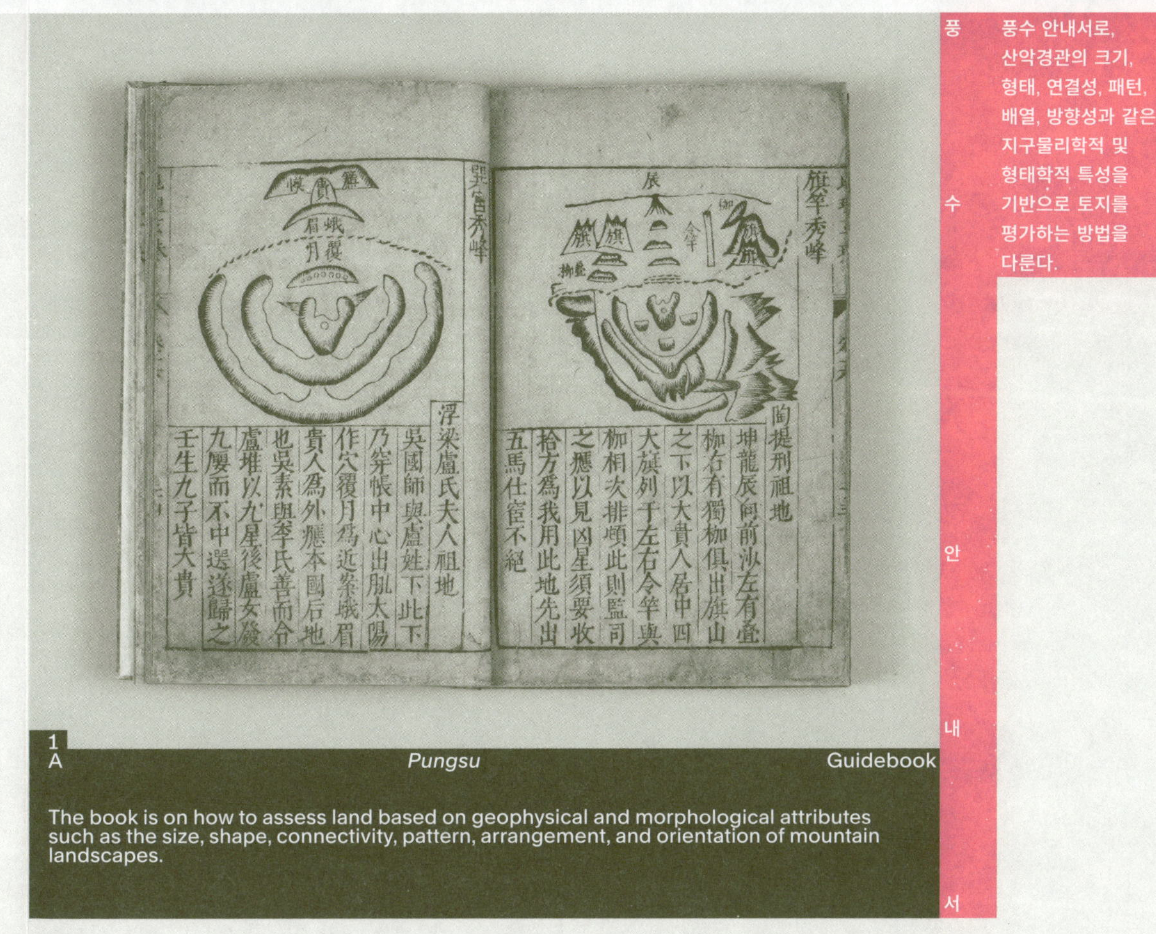

1 A *Pungsu* Guidebook

The book is on how to assess land based on geophysical and morphological attributes such as the size, shape, connectivity, pattern, arrangement, and orientation of mountain landscapes.

풍수 안내서

풍수 안내서로, 산악경관의 크기, 형태, 연결성, 패턴, 배열, 방향성과 같은 지구물리학적 및 형태학적 특성을 기반으로 토지를 평가하는 방법을 다룬다.

To understand the formation and evolution of urban voids nowadays, we need to go back to historical records, looking at the tangible and intangible forces that shaped the city. Consulting these old topographical maps allows us to sketch out a history of the mini-mountains themselves, demonstrating the succession of layers that have gradually framed the remaining unbuilt pockets. These depictions reveal a back-and-forth dynamic of opposing forces: a tide of urbanisation and a resisting breath from the top, or perhaps the spirit of the mountains themselves playing a role in this interplay?

Foundation of the City

From the foundation of the capital city of Seoul, 600 years ago, the presence of the four major inner mountains: Bugaksan 북악산 (342 m), Inwangsan 인왕산 (338.2 m), Namsan 남산 (262 m), and Naksan 낙산 (124 m) and four outer mountains: Bukhansan 북한산 (836.5 m), Deogyangsan 덕양산 (125 m), Gwanaksan 관악산 (632 m), and Yongmasan 용마산 (348 m) have played a significant role in the city's planning. In 1394, during the reign of King Taejo (r. 1392–1398) of the Joseon Dynasty, the area known as Hanyang (referring to the land south of "Hansan," or Bukhansan) officially became the capital of Korea.

To select the site, the Joseon Dynasty followed the system of Korean geomantic principles called *Pungsu* 풍수. *Pungsu* is a form of divination based on topography, which flourished during the Joseon Dynasty. The mythical father of the Korean *Pungsu*, Doseon (827–898), is said to have learned these theories from a mountain spirit, hence the emphasis on mountains. This system evaluates various features of land, mountains, and water, and connects them to human fortune and misfortune. The goals of this system were divided into two categories: creating space for the living, and for the dead. The influence of *Pungsu* can be observed at every level of city planning. It defines the location of the royal palace and commoner dwelling places, as well as the orientation of houses and burial grounds.

도시의 토대

600년 전부터 서울이라는 수도의 토대가 되어준 북악산(342 m), 인왕산(338.2 m), 남산(262 m), 낙산(124 m), 그리고 외사산인 북한산(836.5 m), 덕양산(125 m), 관악산(632 m), 용마산(348 m)은 서울의 도시계획에서 의미 있는 역할을 담당해 왔다. '한양(漢陽, 한산 또는 북한산의 남쪽 땅 지칭)'으로 불렸던 지금의 서울은 조선 태조(재위 1392–1398) 시기였던 1394년에 정식으로 조선의 수도로 지정되었다.

조선왕조는 수도를 선정하면서 한국의 풍수지리(風水地理) 원칙을 따랐다. 조선시대에 널리 퍼진 풍수는 지형에 기반을 둔 점술이다. 한국 풍수의 아버지로 불리는 신화적 인물 도선대사(道詵, 827–898)는 산신령으로부터 풍수 원칙을 배웠다고 하며, 그러므로 당연히 산을 중요하게 여긴다. 풍수는 땅과 산, 물의 다양한 특성을 평가해 인간의 길흉화복과 연결한다. 풍수의 목적은 산 사람을 위한 공간과 죽은 사람을 위한 공간을 조성하는 것, 이렇게 두 가지로 구분된다. 풍수는 도시계획의 모든 단계에 영향을 미치며 궁과 평민 주거지의 위치, 집과 묘지의 방향을 정하는 기준이었다.

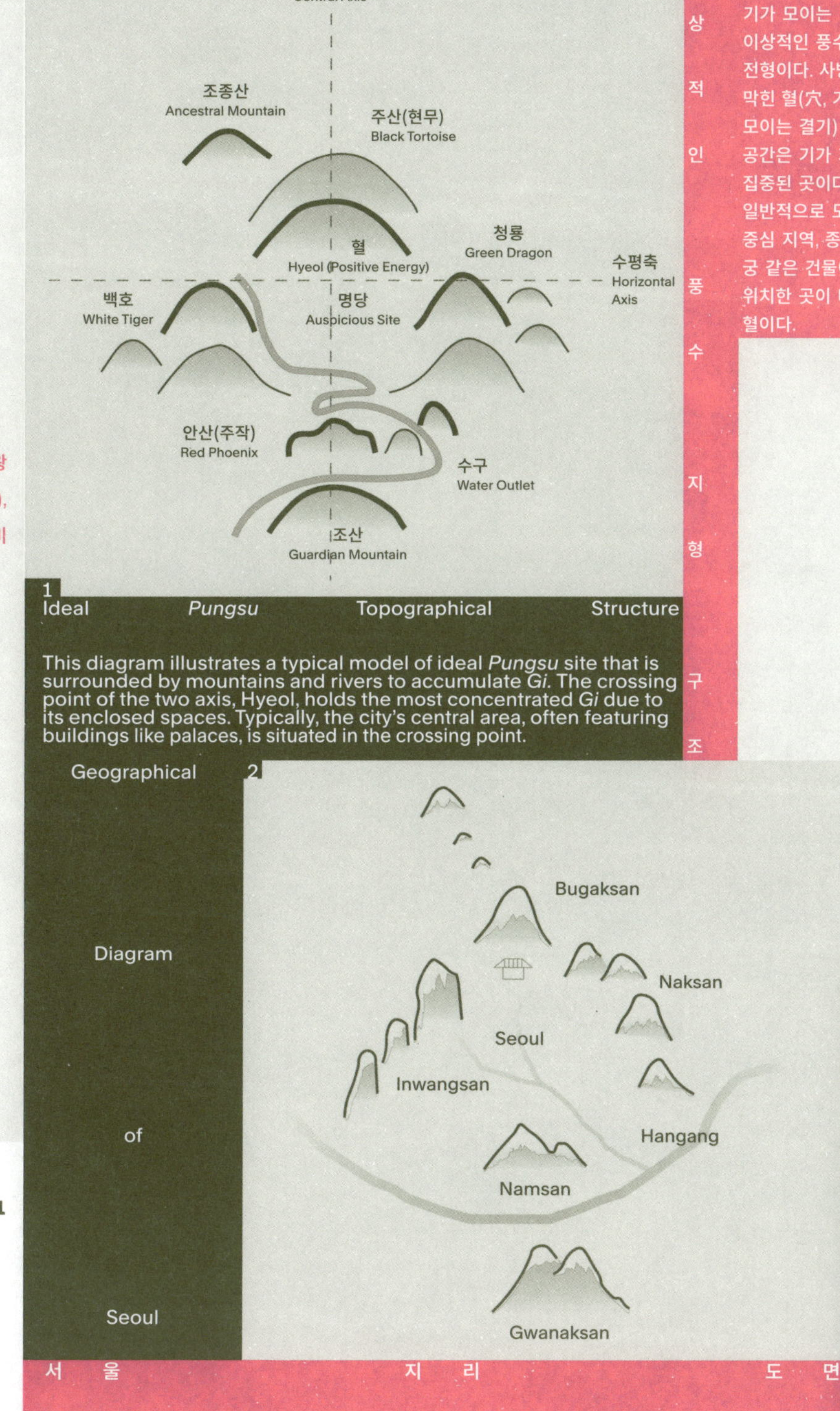

1 Ideal *Pungsu* Topographical Structure

이상적인 풍수 지형 구조

This diagram illustrates a typical model of ideal *Pungsu* site that is surrounded by mountains and rivers to accumulate *Gi*. The crossing point of the two axis, Hyeol, holds the most concentrated *Gi* due to its enclosed spaces. Typically, the city's central area, often featuring buildings like palaces, is situated in the crossing point.

이 도면은 산과 강으로 둘러싸여 기가 모이는 이상적인 풍수 터의 전형이다. 사방이 막힌 혈(穴, 기가 모이는 곁기) 공간은 기가 가장 집중된 곳이다. 일반적으로 도시의 중심 지역, 종종 궁 같은 건물이 위치한 곳이 바로 혈이다.

2 Geographical Diagram of Seoul

서울 지리 도면

3 Panoramic View of Seoul at the Beginning of the 20th Century

20세기 초, 서울 전경

In *Pungsu*, all materials and energy are derived from *Gi* (氣). Based on an area's topography, *Gi* can become denser or thinner. Humans are said to harmonise with nature through *Gi*. In the past, mountains were thought to "breathe" Earth-*Gi* in and out, with the mountain ranges regarded as Dragon Pulses through which *Gi* flows. Tall mountains were seen as yang, so the buildings within the city were kept short to maintain a yin-yang balance and avoid misfortune. From a climatic perspective, maintaining relatively low building mass facilitates the flow of wind from the mountains. In traditional houses (*Hanok* 한옥), the flow of wind is distributed through an internal courtyard. From both a *Pungsu* and a climatic point of view, the topographical and typological buffers were fundamental for the well-being of the city's inhabitants.

풍수에 따르면 모든 물질과 에너지는 기(氣)에서 나오며, 지형에 따라 기는 세지거나 약해진다. 기를 통해 인간과 자연이 조화를 이룬다고 믿으며, 과거에는 기가 흐르는 용맥으로 생각했던 산맥을 통해 산이 땅의 기운을 '숨'처럼 들이쉬고 내뱉는다고 보았다. 높은 산의 기운을 양기로 보았기 때문에 음양의 균형을 유지하고 액운을 피하고자 도시 건물을 낮게 지었다. 기후적 관점에서 볼 때,상대적으로 높이가 낮은 건축물이 있으면 산에서 불어오는 바람의 흐름이 막히지 않는다. 한옥에서는 바람이 안마당으로 흐르며 분산된다. 풍수와 기후 측면에서 지형적, 형태적으로 완충 기능을 수행했던 공간은 도시에 거주하는 사람들의 웰빙을 고려한 기본 개념이었다.

1
Holistic Landscape of Man and Mountain on Korean Peninsul
한반도 산맥의 계

Black Tortoise
(Peak to the north)
Bugaksan
Inwangsan
Naksan
White Tiger
(Ridge to the west)
❷ Hyeol
(Positive Energy)
Blue Dragon
(Ridge to the east)
Namsan
Red Phoenix
(Good open view to the south)
Hangang
(Water from the river in the front)
Gwanaksan

1
Setting the Capital According to *Pungsu* 풍수에 따른 수도 선정

The *Baekdudaegan* mountain range extends throughout the Korean Peninsula and serves as the boundary between China and North Korea. It consists of peaks like *Baekdu*, considered the grandfather, *Halla*, the grandmother, and other peaks in between, creating a unified family of mountains. Acting as a central axis, the *Baekdudaegan* channels natural energies along its length, distributing them through its secondary ridges, valleys, and ultimately into the lands, crops, and inhabitants of Korea, shaping a holistic connection between humans and the mountains.

백두대간 산맥은 한반도를 가로지르며 중국과 북한의 경계선을 이룬다. 이 산맥은 할아버지로 여겨지는 백두산, 할머니인 한라산, 그리고 그 사이의 여러 다른 봉우리들로 연결되며, 하나의 통합된 산악군을 형성한다. 백두대간은 한반도의 중심축으로서 자연의 에너지를 전달한다. 이차 능선과 계곡을 따라 한국의 땅과 농작물, 주민들에게 다다름으로써 인간과 산 사이의 유기적인 연결 고리를 형성한다.

Mini-Mountains	Significant Mountain Ridge	Gyeongbokgung	Changdeokgung
Seoul Old City Wall	❶ The Mountain-range-spine of the Korean Peninsula continuously feed essential energy throughout the nation.	❷ An auspicious site on the slope of a mountain where the energies of Heaven and Earth are very well-balanced.	

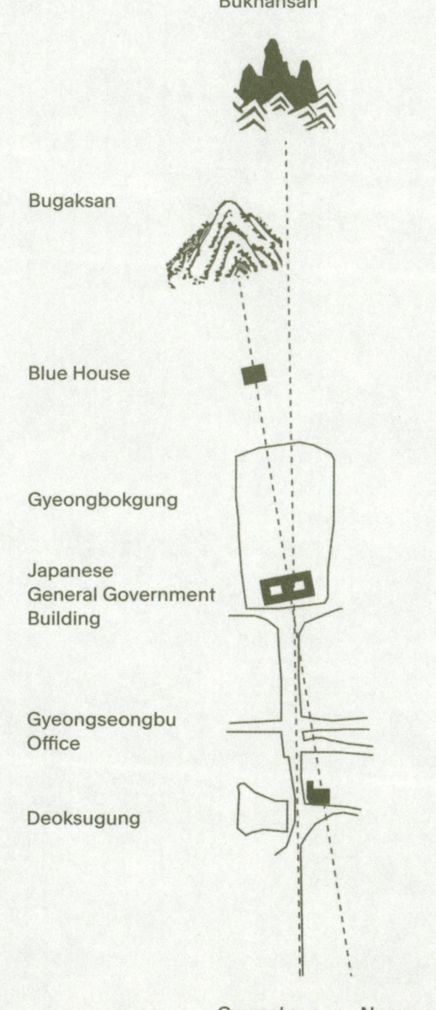

1
Shifting of Axis 축의 변화

Bugaksan resembled the character 大, while the Japanese General Government Building layout was designed in the shape of the character 日, and the Gyeongseong building bore a resemblance to the character 本. To emphasize a new axis, the Japanese General Government Building was purposely oriented 3.5 degrees away from the original alignment. When viewed from above, the characters 大 日 本, meaning Great Japan, were visibly formed.

북악산은 '대(大)' 자를 닮았고, 조선총독부 청사의 배치는 '일(日)' 자 형태로 설계되었으며, 경성부 청사는 '본(本)' 자와 유사한 모습이다. 새로운 축을 강조하기 위해 조선총독부 청사는 원래 정렬에서 3.5도 틀어지도록 의도적으로 조정되었다. 위에서 바라보면 '大日本' 자가 시각적으로 형성되어 '대일본(Great Japan)'이라는 의미를 드러낸다.

Beyond the climatic advantage of locating the city within a circular mountain ridge, the dramatic landscape of peaks and valleys provides a natural defence barrier. The name of (3) 봉화산 *Bonghwasan* or "beacon fire," is testament to the fact that the mountain was used as a beacon station, using fire or smoke to send signals to the city.

Despite these auspicious conditions, the city experienced dramatic events during the Japanese Colonial Period of Korea (1910–1945). To assert its power, Japan sought to destroy the patriotic will and national vitality of the Korean people through acts commonly called "*Pungsu* terrorism." Aware of the importance of mountains and their spirits, the Japanese army drove iron stakes into Seoul's auspicious mountains to disrupt the flow of *Pungsu-jiri* in Korea. Another notorious act of *Pungsu* terrorism occurred in 1926, when the Japanese altered the original axis that visually connected the traditional seat of the Joseon dynasty, Gyeongbokgung Palace, to Bukhansan in the north and Gwanaksan in the south. By rotating the orientation of the Japanese General Government Building (조선총독부 청사) in front of the Palace by 3.5 degrees, a new axis connecting the Bugaksan in the north to the Namsan in the south was established. This arrangement, which had symbolic implications, was intended to undermine the spiritual energy of the capital city by diminishing the significance of the Palace and the spirit of Joseon.

After the Japanese Colonial Period ended at the fall of the Japanese Empire in 1945, the Seoul government sought to restore its identity by extracting the iron stakes from the mountains and reestablishing the original axis. These efforts demonstrate that *Pungsu* and politics were still closely linked in the city's reconstruction during the twentieth century.

둘러싼 산등성이 안에 도시가 위치해 얻을 수 있는 기후상 이점 외에도, 산봉우리와 계곡의 변화무쌍한 지형은 자연 방어벽을 제공한다. **봉화산** (3) 이라는 이름을 보면, 도시로 신호를 전달하기 위해 불이나 연기를 사용했던 봉수대로 사용되었다는 사실을 알 수 있다.

서울은 이런 명당에 위치했지만 일제강점기에(1910–1945)에 비극적인 사건들을 겪었다. 일본은 힘을 과시하기 위해 '풍수 저해'로 볼 수 있는 행위를 함으로써 한국인의 애국심과 국가의 정기를 파괴하고자 했다. 산과 거기에 깃든 정기의 중요성을 알았던 일본군은 한국 풍수지리의 흐름을 방해하고자 서울 명산에 쇠말뚝을 박았다. 1926년에는 북쪽의 북한산과 남쪽의 관악산을 연결하는 축선과 일치하도록 자리 잡은 조선왕조의 전통 경복궁 축선을 변경하는 끔찍한 풍수 저해 행위를 저질렀다. 궁 앞에 지은 조선총독부 청사의 방위를 3.5도 비틀어 북쪽의 북악산과 남쪽의 남산을 연결하는 새로운 축선을 만들었다. 이 행위는 중요한 상징 배치를 변경함으로써 궁의 위상과 조선의 정기를 약화해 수도의 기를 꺾으려는 의도였다.

1945년 일본 제국이 멸망하면서 일제강점기가 끝났고, 서울시는 산에 박힌 쇠말뚝을 제거하고 경복궁의 원래 축선을 복원해 정체성을 회복하고자 했다. 20세기 도시 재건 과정에서도 풍수와 정치가 여전히 밀접하게 연결되어 있었다는 말이다.

1 Hanyangdoseongdo, "Map of the Capital," 1770

한양도성도, 1770

The study of old maps revealed that, as far back as historical records show, mountains were consistently represented as emblematic features and landmarks of Seoul. Maps from the eighteenth century show the city of Hanyangdo, as Seoul was previously called, framed by protective rings of mountains, which were connected and protected by a defensive wall. Within the built city centre, the palaces and shrines appear on the map as buffer zones, microcosms of wooden pavilions amid trees and open spaces. Between the official administrative enclaves and the mountains, the remaining flat area within the city was left for the civilians. Outside the city wall, the mountainous landscape inspired poets and artists, and was also a favourable territory for temples, local shrines, altars (to the local religion of Confucianism), and royal tombs. In the map of Yangcheon, for example, we can locate the mini-mountains of 24 개화산 *Gaehwasan* and its related Yaksasa Temple.

고지도(古地圖)

역사 기록에서도 찾아볼 수 있듯, 고지도에 표시된 서울의 상징적인 특징과 랜드마크는 일관성 있게 산이다. 18세기 지도를 보면 한양도성이 방어벽처럼 연결 고리를 형성한 여러 산에 둘러싸여 보호받았다는 것을 알 수 있다. 번화한 도시 중심부에 자리 잡은 궁과 사당은 지도를 들여다보면 나무와 개방된 공간 속에 자리한 정자의 작은 세계, 일종의 완충 구역으로 나타난다. 공식 행정구역과 산들 사이의 평지는 민간인의 몫이었다. 성벽 밖의 산악 지형은 시인과 예술가에게 영감을 주었고, 또한 사원과 지역 사당, 지역 유교 제단, 왕릉 등의 위치로도 선호되었다. 예컨대 양천 지역 지도에서는 **개화산** 24 의 작은 산들과 산사인 약사사(藥師寺)를 찾아볼 수 있다.

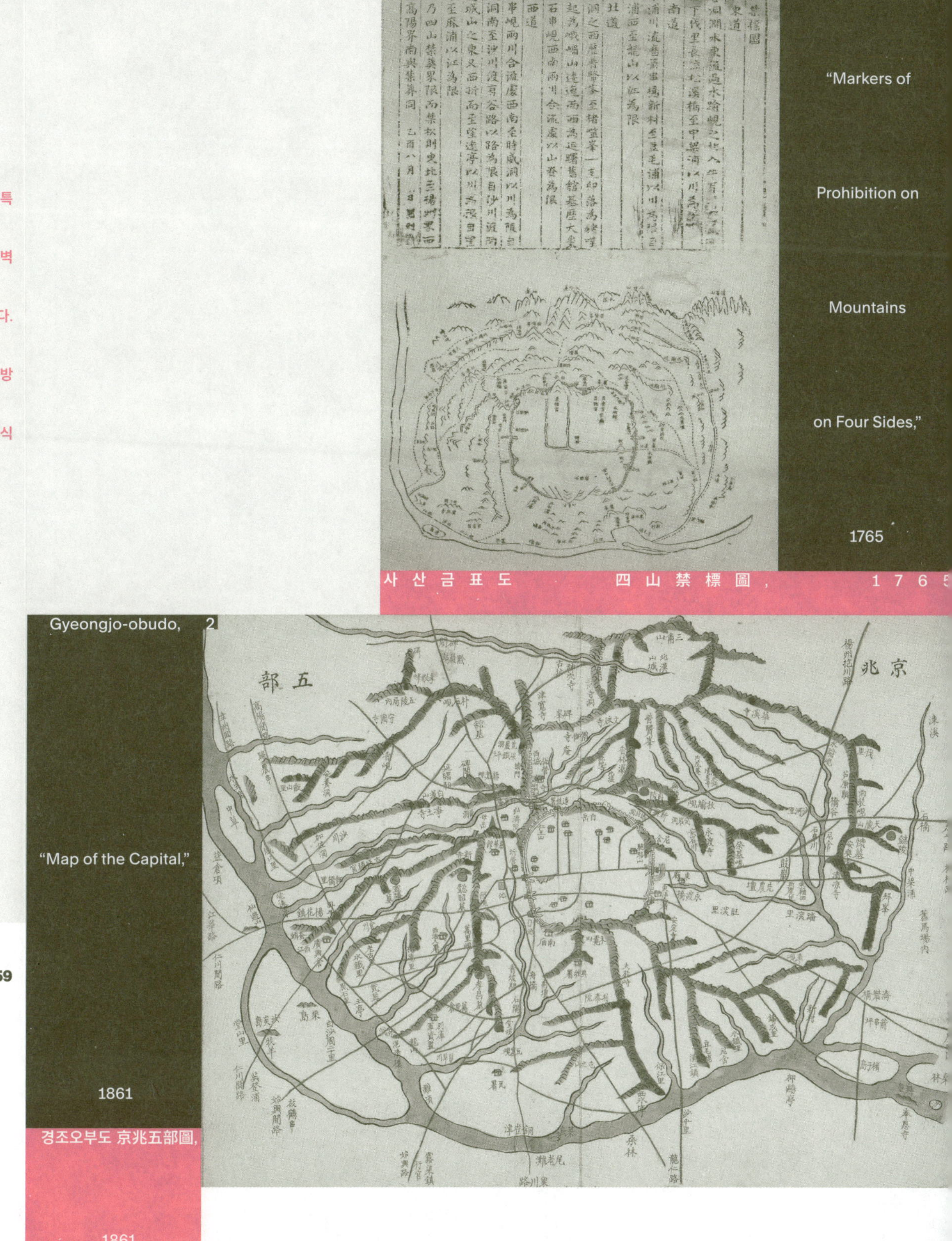

1 Sasangeumpyodo, "Markers of Prohibition on Mountains on Four Sides," 1765

사산금표도 四山禁標圖, 1765

2 Gyeongjo-obudo, "Map of the Capital," 1861

경조오부도 京兆五部圖, 1861

The city was never solely contained within the wall, and connections with the suburbs provided vital resources. The nearby settlement of Seongjeosimni and its surrounding mountains, for instance, were an important source of wood for the city. The population within the capital expanded during the late Joseon Dynasty. In 1765, the map Sasangeumpyodo (사산금표도 四山禁標圖, "Markers of Prohibition on Mountains on Four Sides") was produced and distributed to the private sector to prevent indiscriminate logging and expansion of burial grounds near the capital city. The map clearly shows boundaries where logging of pine trees was prohibited within the capital city walls and within Seongjeosimni, with markers to prevent the creation of funeral sites.

A century later, the Gyeongjo-obudo ("Map of the Capital") produced by Kim Jeong-ho in the 1840s and published in 1861, included in the large scale map called Daedongyeojido, shows the old city periphery. This map represents the city surrounded by mountains, leaving the centre largely empty. Only a few houses are depicted, and no writing accompanies them. However, the names of mountains, temples, bridges, and offices in outer Seoul are recorded in detail.

한양은 성벽 안에 고립된 도시가 아니라, 중요한 자원을 제공하는 외곽부와 늘 연결된 수도였다. 예를 들어 성저십리(城底十里) 내외에 위치한 산 인근의 거주지는 수도에 목재를 공급하는 중요한 지역이었다. 조선 후기에 수도 인구가 점차 증가하면서, 1765년에는 무분별한 벌목과 수도 근처에 묘지가 더 세워지는 것을 막기 위한 지도 「사산금표도(四山禁標圖)」를 만들어 민간에 배포했다. 지도에는 수도 성벽과 성저십리 내에서 소나무 벌목과 묫자리 조성을 금지하는 구역과 경계선이 분명하게 표시되었다.

그로부터 1세기가 지난 후, 1840년대 김정호(金正浩)가 제작해 1861년 발간한 대축척 『대동여지도(大東輿地圖)』의 「경조오부도(京兆五部圖)」는 옛 수도의 경계선을 표시했다. 산으로 둘러싸인 도시와 대부분 빈 공간인 중심부를 볼 수 있는데, 중심부에는 설명하는 글 없이 몇 채 안 되는 집만 그려져 있다. 하지만 서울 외곽의 산과 사찰, 다리, 관청 이름은 자세히 기록되었다.

1 Map of Gyeongseongdo, 1922
경성도, 1922

The 1922 map published during the Japanese Colonial Period shows a shift in representations of the city. Drawn from above and according to orthographic principles, it was the most accurate and largest map of Seoul so far. The first measurements were taken in 1915, and the second revised set of measurements was made in 1921, with the final map published in July 1922. This document provides a detailed survey of formal buildings (including the recently built Japanese General Government Building and topographical records. It encompasses a wider area from the city centre, showing the first movement of city expansion towards the Hangang River area. As a good portion of the map includes areas that remained unbuilt, it provides valuable topographical information about previous landforms and occupation of some mini-mountains today. It includes the existing locations of several mountains: 2 배봉산 *Baebongsan*, 5 응봉산 *Eungbongsan*, 9 노고산 *Nogosan*, 11 서달산 *Seodalsan*, and even

일제강점기에 발간된 1922년 지도는 수도를 다르게 보여준다. 위에서 바라본 정사도법으로 제작된 이 지도는 당시 서울을 가장 정확하게 묘사한 대형 지도였다. 첫 측정은 1915년에 진행되었고, 1921년에 2차 측정을 거친 후, 최종본은 1922년 7월에 발간되었다. 이 지도는 정부 기관 건물들(당시 지어진 조선총독부 청사 포함)의 정보와 지형 기록을 자세히 제공한다. 도시 중심부보다 더 넓은 지역을 담은 이 지도는 도시가 한강 지대로 확장해 변화하던 초기 모습을 보여준다. 지도의 넓은 면적에는 건물이 들어서지 않은 빈 공간이 포함되며, 과거 지형과 현재 작은 산이 차지하는 공간을 비교할 수 있는 소중한 지형 정보다. 배봉산 2, 응봉산 5, 노고산 9, 서달산 11의 기존 위치와 더불어 천장산 1을 포함한 사찰과 무덤의 위치도 표시되었다.

1 Kyongsong or Seoul (Keijo), 1946

경성 또는 서울(게이조), 1946

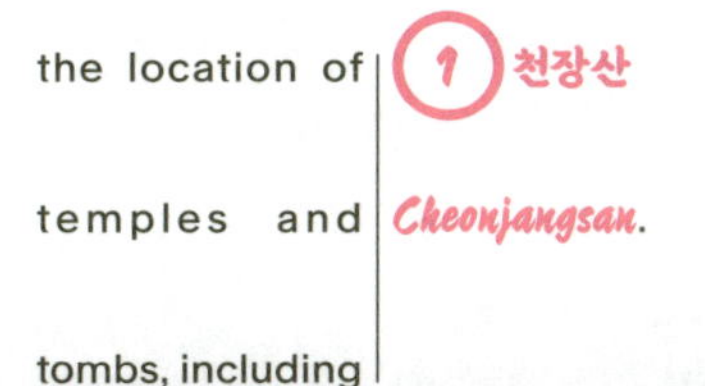
the location of temples and tombs, including 1 천장산 *Cheonjangsan*.

In 1946, the Kyongsong or Seoul (Keijo) map was compiled and published as one of the military topographic maps and related products required by the Armed Forces of the United States, who administered what is today South Korea in the postwar period. This map highlights the densely built-up areas of the old city, expanding to the south-west (shown in yellow) and less densely built areas (shown in hatched yellow). Besides providing valuable topographical information, the map also includes detailed information on various natural features such as woodland, grassland, or rice fields. This helps to better understand the original context of the mini-mountains, including those such as 5 응봉산 *Eungbongsan*, 6 와우산 *Wausan*, 7 궁동 *Gungdong*, 8 성산 *Seongsan*, 9 노고산 *Nogosan*, and 13 용마산 *Yongmasan*.

1946년의 지도 「경성 또는 서울(게이조) (Kyongsong or Seoul (Keijo)」은 전쟁 후 오늘날 남한이 된 지역을 보호하던 미국군에 필요했던 군사 지형도와 관련 자료를 통합해 제작되었다. 이 지도는 건물이 밀집한 구시가지 지역을 강조하면서, 남서쪽으로 확장된 지역(노란색으로 표기)과 밀집도가 낮은 지역(노란색 줄무늬로 표기)을 보여준다. 값진 지형 정보를 제공하는 이 지도는 삼림지대, 초원, 논밭 등 다양한 자연 지형 역시 자세히 표시한다. 또한, **응봉산** 5, 와우산 6, 궁동 7, 성산 8, 노고산 9, 용마산 13 같은 작은 산의 원래 환경을 파악할 수 있다.

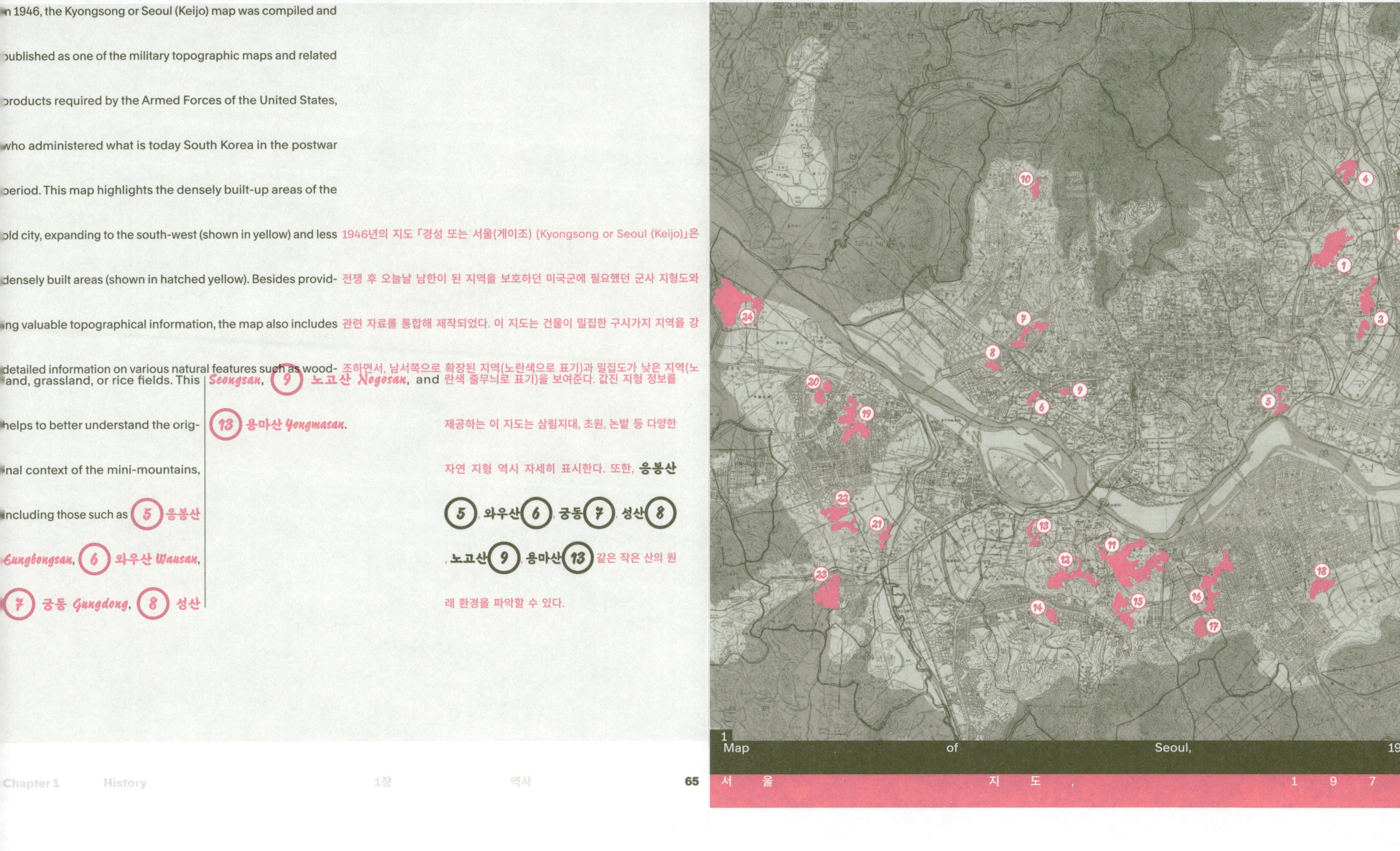

1 Map of Seoul, 1975
서울 지도, 1975

From 1930 onwards, maps not only reflected the city in its current state but also served as a tool to communicate its future strategic development. Following the Korean War in 1953, Seoul underwent rapid urbanisation to accommodate a population that grew from 1.6 million in 1955 to 10.6 million in 1990. To cope with this population growth and relieve congestion north of the Hangang River, the city started to develop new centres south of the river, first to the south-west and later to the south-east, such as the Gangnam District. By comparison, the expansion of the Seoul metropolitan area south of the Han River represents a drastic change. In less than fifty years, the original rice fields and wetlands were transformed by urban sprawl, with buildings now occupying the slopes and framing residual pockets of nature on the hilltops.

When looking at the city from above, it is difficult to believe that the homogeneous distribution of building mass across the Seoul Metropolitan Area occurred within just half a century. However, examining successive development layers around the mini-mountains offers deeper insights into the profound morphological modifications the city has endured.

1930년 이후 지도는 도시의 현황뿐 아니라 향후 진행할 전략적 개발을 알리는 도구 역할도 했다. 1953년 한국전쟁 이후 서울은 1955년 160만 명에서 1990년 1,060만 명으로 급증한 인구를 수용하기 위해 빠른 도시화 과정을 겪었다. 이렇게 급속한 인구 증가에 대응하고 강북 지역의 혼잡을 완화하기 위해, 서울은 처음에는 한강의 남서쪽으로 확장하기 시작하고 나중에는 강남구 같은 남동쪽으로 확장하면서 강남 지역에 새로운 중심지를 개발했다. 수도권의 강남 쪽 확장은 상대적으로 극적인 변화를 보여준다. 50년도 채 지나지 않아 원래 논밭과 습지였던 곳이 완전히 변형되고, 지금은 비탈길에 들어선 건물들이 언덕 상부에 남아 있는 자연 공간을 둘러싸고 있는 형태로 변모했다.

상공에서 서울을 내려다보면 서울 수도권 지역에 균등하게 들어선 건물의 배치가 불과 반세기 만에 생긴 변화라는 사실을 믿기 어렵다. 하지만 작은 산 주변에서 잇따라 진행된 개발 층위를 살펴보면 서울의 형태적 변형이 얼마나 엄청난 규모로 이루어졌는지 알 수 있다.

The Gradual Framing of Voids

At a smaller scale, the fluctuation of the city's morphology can be observed by examining the transformation of the mini-mountains. In the 1922 map, the mini-mountains didn't exist as "voids" (un-built areas framed by buildings). Each hill appearing in this book was simply immersed within the ripples of the terrain, connected to and expanding from the main mountains.

Gradually, after the 1922 maps, we start to see the first built occupation of the terrain surrounding the hills. In particular, there is a shift between the maps from 1953 and the 1960s, to those from the 1970s. Starting from 1970, the maps indicate low-rise built areas with a red surface, framing the mountains, while major buildings such as official institutions, universities, and apartment blocks are shown in black (1) 천장산 *Cheonjangsan*. This transformation happened within a decade for (6) 와우산 *Wausan*, but slightly slower for (7) 궁동 *Gungdong*. In 1977, there were still patches of undeveloped land, which turned to compact red by 1986. In the south of the river, (20) 우장산 *Ujangsan* and (18) 매봉산 *Maebongsan* show a similar process, except that it starts from a less dense built context where the houses can still be read as independent units identify the small units, but gradually they are replaced by larger blocks.

At the boundary of the metropolitan area, the transformation surrounding (24) 개화산 *Gaehwasan* goes from sparse small units and fields until about 1970. From the 1980s onwards, we can observe a systematic organisation of neighbourhoods (either small buildings or blocks). However, the major transformation is the densification of transport infrastructure (primarily expressways due to its proximity to Gimpo Airport).

보이드의 점진적인 형성

작은 산의 변형 과정을 자세히 보면 도시 형태가 어떻게 변했는지 파악할 수 있다. 1922년 지도에서 작은 산은 '보이드(건물들로 둘러싸인 미개발 지역)'로 표시되지 않았다. 이 책에서 묘사하는 모든 작은 산은 산에서 뻗어 나가는 물결 같은 지형의 일부였을 뿐이다.

그러나 그 후, 작은 산을 둘러싼 지형에 건물들이 점차 들어서는 모습을 지도에서 확인할 수 있다. 특히 1953년과 1960년대 지도, 그리고 1970년대 지도에서 변화가 가장 두드러진다. 1970년부터 산을 둘러싸고 높이가 낮은 건물이 들어선 구역이 빨간색으로, 관공서·대학교·아파트 단지 등 큰 건물이 들어선 구역이 검은색으로(**천장산** (1)) 표시되었다. **와우산** (6)은 이렇게 변화하는 데 채 10년도 걸리지 않았지만 **궁동** (7)의 경우에는 변화 속도가 더디었다. 1977년에는 아직 개발되지 않은 토지가 남아 있었지만 1986년에는 많은 지역이 밀집된 빨간색으로 바뀌었다. 한강 이남의 **우장산** (20)과 **매봉산** (18)도 비슷한 양상을 보였으나 초기에는 주택이 작은 독립 단위로 표시되는 한산한 지역이었다가 점차 큰 건물들로 교체되었다는 점이 다르다.

수도권 경계 지역에서 **개화산** (24)을 둘러싼 지역은 1970년까지 밭과 드문드문 들어선 작은 건물 표시로 가득했으나 1980년대 이후 체계적으로 구조화된 지역 단위(작은 건물 또는 단지)로 변화한다. 하지만 두드러지는 변화는 교통 기반 시설이 집중되면서(김포공항과 가까워 고속도로가 발달하면서) 생겼다.

The shrinking of unbuilt areas toward the mountain's top is a phenomenon that is relatively to be expected as the city expands. However, the drastic changes in the configuration of the built environment in 5 응봉산 *Eungbongsan* and 12 국사봉 *Guksabong* between the 1970s and 2000s were less predictable. Until the 1950s, 12 국사봉 *Guksabong* remained largely an untouched natural area (there were even plans for it to be kept as a nature park—see Chapter 3). Yet from the 1960s onwards, patches of built mass (represented in red) started to cover the mountain's ridge. By the 1970s, the mountain's ground was totally fragmented, with very few patches of unbuilt land remaining at its top. This encroachment on its empty spaces continued until the 1990s, and it's only in the 2000s that some pockets of voids reopened. Nowadays, the pockets have reconnected, and trees have regrown over the ridge. The gradual decolonisation of the mountain by small built units implied a transformation into apartment blocks.

In the case of 5 응봉산 *Eungbongsan* in 1946, we have indications that the top of the northern part and the north of the southwest parts were forested, while the lower connecting part was mostly undeveloped. By the 1980s, the north part of the mountain had been completely covered by housing settlements but today, the mountain's summit has also returned back to its earlier unbuilt state, whereas the surrounding are framed by apartment blocks.

To better understand the distribution of solids and voids within the new Seoul Metropolitan Area, we need to consider the macro-scale and understand the shift in the city's vision regarding public spaces and housing policies. Looking more closely at the shift in the ways of depicting the city (Chapter 3) indicates that the process wasn't entirely natural: instead it's a political vision and a collective effort to realise it that allows the restoration of nature to occur.

산 정상 주변의 미개발 지역이 줄어드는 것은 도시가 확장할 때 예상할 수 있는 현상이다. 하지만 1970년대와 2000년대 사이에 **응봉산** 5 과 **국사봉** 12 이 갑자기 밀집 지역으로 전환된 모습은 예상치 못한 변화였다. 1950년대까지 **국사봉** 12 은 거의 훼손되지 않은 자연 생태 공간이었다(원래 자연 생태 공간으로 보존할 계획도 있었다—3장 참조). 하지만 1960년대 이후부터 건물 단지(빨간색 표시)가 국사봉의 산등성이를 덮기 시작했다. 1970년대에는 대부분의 지면이 건물 단지로 완전히 덮여 산 정상부 주변에 미개발 공간은 거의 남지 않게 되었다. 빈 공간을 잠식하던 개발은 1990년대까지 계속되다가, 2000년대에 들어서야 일부 빈 공간이 다시 조성되기 시작했다. 요즘은 빈 공간들이 서로 연결되어 산등성이가 다시 나무로 덮이고 산에 표시되었던 작은 건물 단위는 점차 아파트 단지로 대체되고 있다.

응봉산 5 의 경우에는 1946년 산의 북쪽 정상과 남서쪽 북부에 숲이 남아 있고 산 아래쪽에 연결된 지역은 미개발 상태였다는 기록이 있다. 1980년대에는 산 북쪽이 주택가로 완전히 덮였지만, 지금은 정상부도 예전처럼 빈 공간으로 복원되었고 그 주변을 아파트 단지가 둘러쌌다.

새로운 서울 수도권 지역 내 보이드와 솔리드(건물들이 조성된 공간)의 분포를 잘 파악하려면 거시적인 관점에서 규모를 살피고, 공용 공간과 주택 정책에 대한 도시 비전의 변화를 이해해야 한다. 도시가 어떻게 변했는지 자세히 살펴보면(3장) 그 변화가 자연히 이루어진 것이 아니라 자연환경을 복원하려는 정치적 비전과 이를 실현하기 위한 집단적 노력의 결과임을 알 수 있다.

Year	1920	1930	1940	1950	1960
1 천장산 *Cheonjangsan*	1922	1930		1953 1957	
5 응봉산 *Eungbongsan*	1922		1946	1950 1958	1966
6 와우산 *Wausan*		1930	1946	1950	1967
7 궁동 *Gungdong*		1930	1940	1953	1967
12 국사봉 *Guksabong*		1930	1942	1950	1963

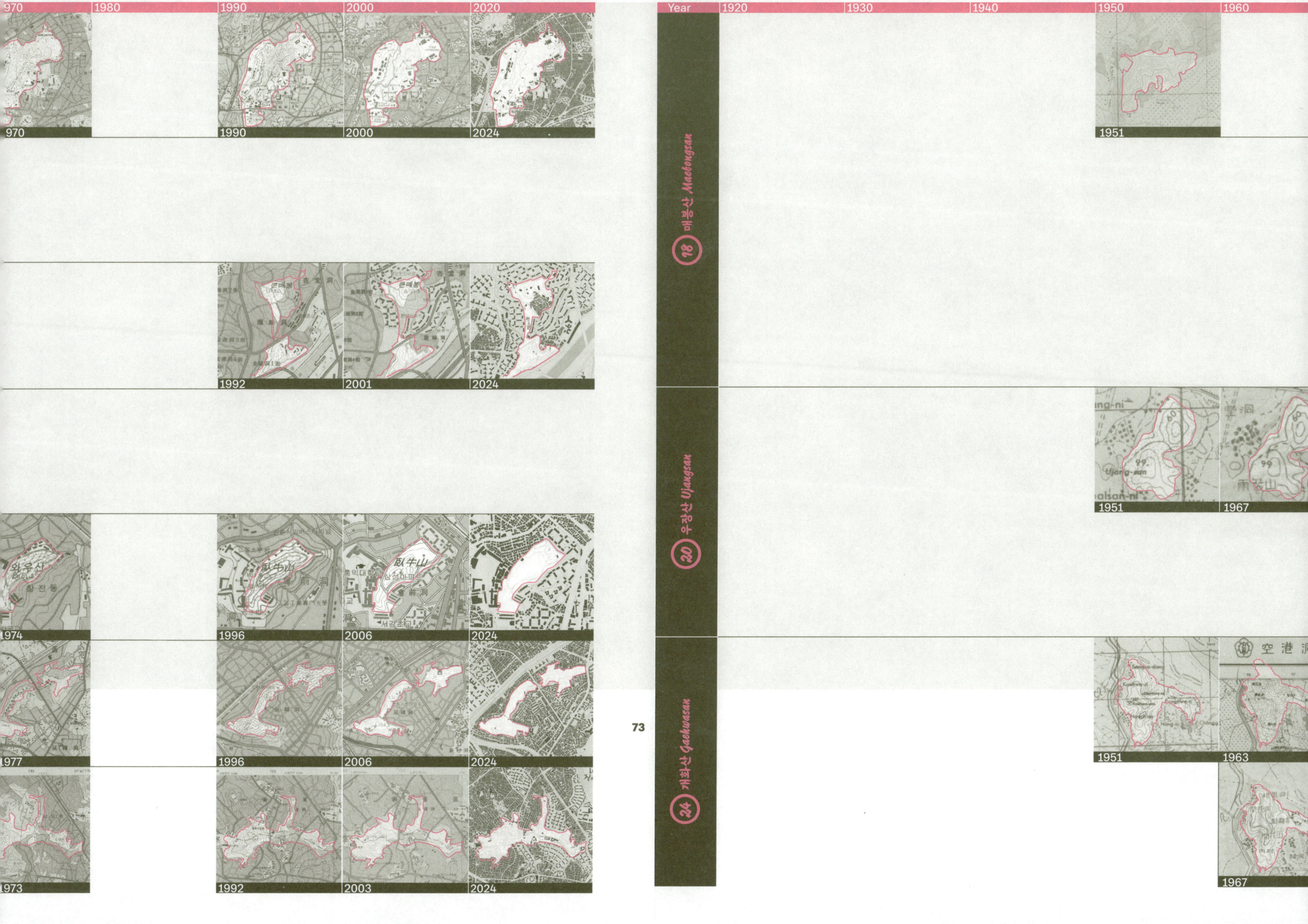

970
1980
1990
2000
2020
970
1990
2000
2024
1992
2001
2024
1974
1996
2006
2024
1977
1996
2006
2024
1973
1992
2003
2024
Year
1920
1930
1940
1950
1960
18 매봉산 Maebongsan
1951
20 우장산 Ujangsan
1951
1967
24 개화산 Gaehwasan
1951
1963
1967

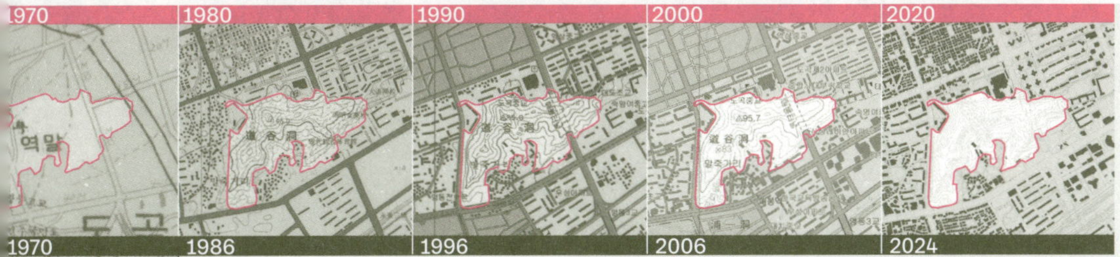
1970
1980
1990
2000
2020
1970
1986
1996
2006
2024

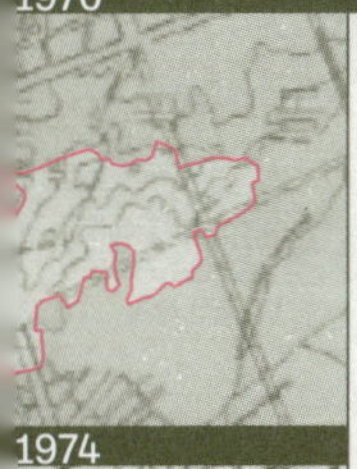
1974

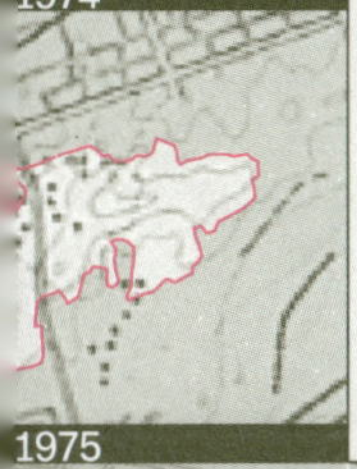
1975

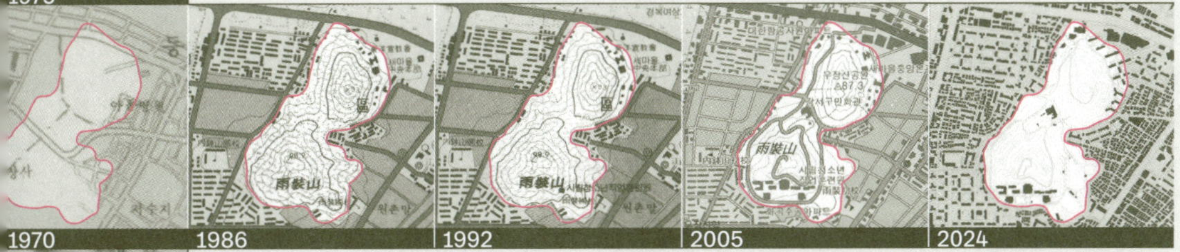
1970
1986
1992
2005
2024

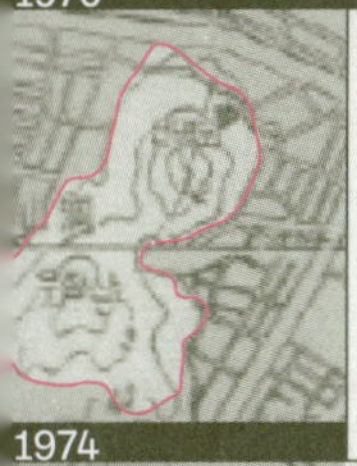
1974

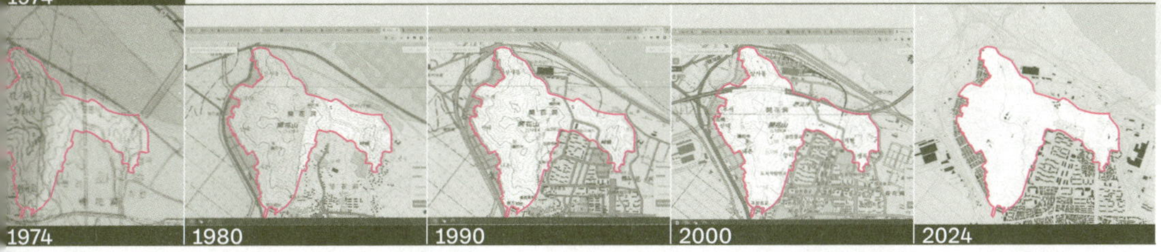
1974
1980
1990
2000
2024

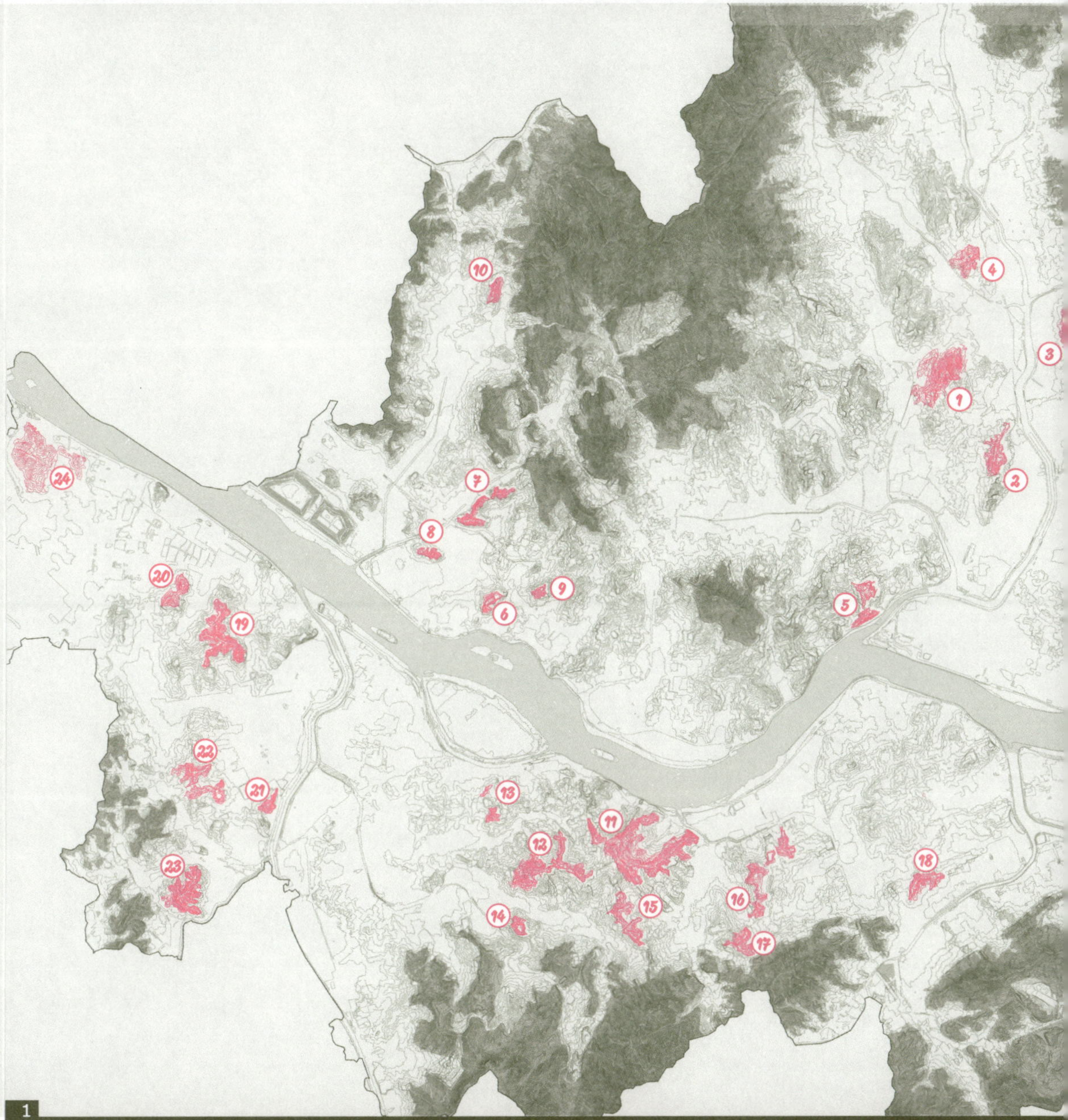
1
2
3
4
5
6
7
8
9
10
11
12
13
14
15
16
17
18
19
20
21
22
23
24

1
Topography

지형

Topography

Topography establishes a sense of distance within the ity; it allows the inhabitants to feel they are escaping the city ithout actually leaving it. When Seoul's contemporary map cleared of its buildings and road network, the remaining ontour lines reveal the turbulent movement of the ground. ompared to cities built on flat terrain with direct and efficient oad networks, cities built on topographically uneven ground eal with obstacles. In the case of Seoul, the city has developed round the mountains, leaving the less accessible peaks as mpty residual spaces. However, when a path leads to the top, usually follows a contour line, contributing to a feeling of reater distance travelled. When straight routes are drawn, uch as stairs, the physical sensation of ascending accentuates the feeling of covering a distance. During a mountain hike, e usually add the notion of effort ("km/effort") to the distance walked ("km"), and this effect is also noticeable when transversing a city.

In Seoul, after leaving a residential neighbourhood and climbing the shaded slopes, one arrives at the top and looks back at the city with a feeling of expansiveness. Yet within a few metres, upon entering a grove, one can feel as though one is entering another world, intensifying the sense of depth. Topography provides flexibility in how space is articulated and creates multiple ways to engage with the landscape.

지형

지형은 도시 안에서 거리감을 형성한다. 주민들에게 실제 도시를 떠나지 않고도 도시를 벗어나는 느낌을 준다. 근대 서울의 지도에서 건물과 도로망을 제거하면 지면의 거친 윤곽이 드러난다. 효율적인 도로망을 갖춘 평지에 조성된 도시에 비해 고르지 않은 지형에 조성된 도시는 여러 장벽을 맞닥뜨려야 한다. 서울은 산을 둘러싸고 도시가 발전했기에 접근하기 어려운 산봉우리가 빈 공간으로 남았다. 산 정상으로 길을 낼 때는 주로 등고선을 따라 조성하기 때문에 더 먼 길을 돌아가는 기분이 든다. 계단처럼 직선으로 길을 내면 높이 올라가는 느낌이 들어 거리감이 더 강조된다. 등산할 때는 이동한 거리(km)에 노력(km/effort)이란 개념을 추가하는데, 이 효과는 도시를 횡단할 때도 유효하다.

서울에서는 주택가를 벗어나 그늘진 비탈길을 따라 오르면 산 정상에서 탁 트인 도시를 내려다보며 숨통이 트이는 기분을 느낄 수 있다. 그 후 몇 미터만 걸어가면 다시 숲속으로 들어서게 되고 깊이감이 강조되어 다른 세상에 들어선 듯하다. 지형을 통해 우리는 공간을 유연하게 감각하고 경관을 다양한 방식으로 바라볼 수 있다.

1
천장산
Cheonjangsan

5
응봉산
Eungbongsan

6
와우산
Wausan

7
궁동
Gungdong

12
국사봉
Guksabong

18
매봉산
Maebongsan

20
우장산
Ujangsan

24
개화산
Gaehwasan

Zooming in to the topographical lines on a map, one can almost read the stories of mini-mountains through the different shapes of its contour lines. Starting from the top, the lines are mostly smooth, indicating relatively untouched ground. On descending, however, one can observe straighter and more angular lines, indicating artificial modification of the ground. A careful reading of these lines can give an indication of the type of buildings present. In (7) 궁동 Gungdong, the serrated lines reflect a neighbourhood of small individual housing units. In (6) 와우산 Wausan, the longer strait lines indicate the terraced ground where some apartment blocks were constructed. In the case of (6) 와우산 Wausan, a finer reading reveals that some lines previously altered to accommodate apartment buildings in the north are gradually smoothing with the recent reconversion of the area into a park (see Chapter 3). Through topographical lines, one can observe that the successive phases of urban development has implied an expanding and congesting movement to the framing of these voids. But this reading of the topography also questions the actual boundaries of these voids. If in the first phase of observation, the limits were defined by the boundary line between the solid and the void, another boundary line could be defined between the natural and artificial contour lines.

지도에서 지형선을 확대해 보면 다양한 윤곽선 형태를 따라 펼쳐지는 작은 산의 이야기를 읽을 수 있다. 산 정상의 선은 대부분 매끄러워서 사람의 손길이 거의 닿지 않은 땅이라는 것을 알 수 있다. 하지만 밑으로 내려올수록 곧고 각진 선이 드러나면서, 지면이 인위적으로 변했다는 것이 확실해진다. 이런 선들을 세심히 살펴보면 어떤 건물이 들어섰는지 파악할 수 있다. **궁동** (7) 에서 톱니 모양 선은 작은 개별 주택을 표시한다. **와우산** (6) 의 길고 곧은 선은 아파트 단지가 조성된 계단식 지반을 가리킨다. **와우산** (6) 을 더 자세히 살펴보면 북쪽에 아파트 단지를 조성하기 위해 변형되었던 일부 선이 최근 단지가 공원으로 복원되면서 점차 완만해지는 걸 알 수 있다(3장 참조). 지형선을 통해 잇따른 도시 개발은 확장과 밀집을 거듭하며 보이드를 형성해 왔음을 알 수 있다. 하지만 지형선은 보이드의 실제 경계에 대한 의문을 불러일으키기도 한다. 처음 관찰 단계에서 보이드(빈 공간)와 솔리드(밀집 공간) 사이의 경계선으로 공간을 분리할 수 있다면, 자연적 윤곽선과 인위적 윤곽선 사이에 존재하는 또 다른 경계선을 정의할 수 있을 것이다.

1 Sanshin in (12) 국사봉 Guksabong

국사봉 (12) 의 산신

Spirits

Sanshin 산신 (Mountain Gods)

When entering some of Seoul's hills, the feeling of istance is amplified by the interlocking of two territories: ne sacred and the profane. Throughout Korean history, people ave believed that the peaks and slopes are spiritually alive, habited by or manifesting mountain spirits known as Sanshin 신 (Mountain Gods). Sanshin are symbols of the relationship etween human beings and the ecology of the mountain where hey live, protecting villages, towns, and the Korean nation s a whole. Mountains are considered sacred sites that can ather yin and yang energies and produce positive energy. They lso represent a cultural ideal of living in balanced harmony vith nature, sharing ideals with Daoism, Korean Shamanism, leo-Confucianism, and Buddhism.

Sanshin are often represented as masculine figures urrounded by tigers. They usually reside at the summit of a mountain, which is considered an ideal place for the ruler of the mountain due to its proximity to the sky. The sunny hillside, sheltered from cold winds, is also an auspicious home for the mountain god. For a deity that resides at the foot of a mountain, an altar or a shrine is set up in a bright spot or along a hiking path at the entrance to a village, serving as a communal shrine for private prayers for a good harvest, healing, or a son.

정기

산신

서울의 산에 들어서면 신성한 영역과 세속적인 영역이 맞물리면서 거리감이 증폭된다. 한국에서는 예로부터 산의 정기가 산신으로 나타나며, 산신이 거주하는 산 정상과 산비탈은 영적인 곳이라고 믿어왔다. 산신은 산의 생태 환경과 그곳에 거주하는 사람들 사이의 관계를 상징하며 마을과 도시, 국가 전체를 보호한다. 산은 음양의 기운을 모아 긍정적인 에너지를 발산하는 신성한 장소로 여겨지며 도교와 한국 무속신앙, 성리학, 불교 사상을 공유하고 자연과 어울려 생존하는 이상적인 문화를 상징한다.

산신은 주로 호랑이에 둘러싸인 남성으로 표현된다. 하늘과 가까워서 다스리기에 이상적인 산 정상에 살며, 찬 바람이 닿지 않는 햇볕 좋은 산비탈은 산신에게도 명당자리이다. 산기슭에 사는 산신을 위해서는 양지바른 곳이나 마을 입구의 등산로를 따라 제단이나 사당을 마련하고, 그곳에서 사람들은 풍년이나 치유, 아들을 얻기를 기원한다.

Sansa 산사 (Mountain Temples)

Articulating the territory and protecting the mountain, small shrines from the Period of the Three Kingdoms (fourth-seventh century AD) tradition serve as guardian spirits under the authority of Sanshin. Bohyeon Sanshingak is an example of a shrine dedicated to worshipping the mountain spirit near Bukhansan in Jongno District, Seoul. It is constructed as a small wooden tile house, housing one god within a space of one kan (間), around 2 m, or less. Petitions can be offered to these spirits for rain and village prosperity. With caring and maternal characteristics, the female mountain spirit can also warn, guide, and protect people entering her territory.

Belief in the mountain spirit has also rooted itself in Buddhist ideology. The idea of respecting the mountain drives the construction of temples in the great mountains in the fifth to sixth centuries. Combining knowledge from India and China, Buddhism in Korea gradually developed its unique composition and layout. The form and composition of the temples adapted to the topography, creating a particular style of mountain temple layout. Building one's home on an auspicious site would also carry prosperity to oneself and one's descendants. During the Joseon Dynasty (1392–1910), when Confucianism became the state ideal, Buddhism suffered harsh oppression. Buddhist temples in the cities were demolished, with only temples in remote mountains surviving. As a result, the Buddhist monks who were once a ruling class were downgraded to a lower class. Generous contributions from powerful elites were no longer available, and the surviving temples faced financial hardships. Inefficient structures like long cloisters disappeared, giving way to more flexible architectural arrangements that suited the irregular, mountainous topography.

산사

삼국시대(4–7세기)의 작은 전통 사
은 산신의 권한을 위임받아 지역을 대표하
산을 보호하는 수호신으로서 기능했다. 보
산신각(普賢山神閣)은 서울시 중구 북한
인근에서 산신을 모시는 사당으로, 2미터
채 안 되는 독립된 공간에 신을 모신다. 신
게는 비가 오기를 바라거나 마을의 번창을
원하며, 모성과 배려심이 강한 여성 산신
영역을 침범하는 사람을 경고하면서도 안
하고 보호하기도 한다.

산신을 믿는 관습은 불교 사상에서 비롯되었다. 산을 받드는 개념은 5–6세기 사이 산에 대규모 사찰을 짓는 데 기여했다. 인도와 중국 사상을 통합한 한국 불교는 점차 독특하게 조직되고 분포됐다. 사찰의 형태와 구성은 지형에 맞추어 변화했고 산사는 고유한 방식으로 지었다. 또한 명당에 집을 지으면 집안이 대대로 번창한다고 믿었다. 조선왕조(1392–1910) 시기에는 유교가 국가 통치 이념이 되면서 불교가 가혹하게 억압받았다. 도시 사찰은 철거되고 산사만 남았으며, 한때 지배계층이었던 승려는 하층민으로 전락했다. 산사는 상류층의 기부가 끊겨 경제적인 어려움을 겪었으며, 자연히 효율성이 떨어지는 긴 회랑의 구조가 사라지고 산악 지형에 적합한 유한 건축 구조로 바뀌었다.

Despite these challenges, mountain temples never lost support among common people. Buddhism continued to be practised and gained popularity by incorporating elements from Confucianism and Shamanism. The monastic communities gained socio-political support and financial backing from elite patrons. Although new temples were generally smaller, their numbers increased to accommodate various schools of Buddhism. The more organic temple layout on sloping terrain created a distinct tradition of Buddhist architecture in Korea. The relationship between the buildings and their orientation to natural topography are essential characteristics of this architectural style. In contrast to geometric layouts, this type of architecture is defined by its relationship with the land rather than by rigid structure.

Today, there are about 45 mountain temples located within the Seoul Metropolitan Area, with several situated on the mini-mountains, including on 1 천장산 *Cheonjangsan*, 24 개화산 *Gaehwasan*, 19 봉제산 *Bongjesan*, 12 국사봉 *Guksabong*, and 11 서달산 *Seodalsan*.

Although the temples have faced tumultuous times throughout history, they have always benefited from their natural surroundings. Unlike churches that represent a focal point for the community, temples are to be discovered after crossing a series of invisible boundaries. It is said that the path leading to the temple is equally important as the visit to the temple itself, reinforcing the idea that the journey has intrinsic value. In the mountains, the distance from the secular world is increased. Paths follow contour lines, cross vegetation filters, and articulate along views, deepening the feeling of distance and isolation. Mountain temples are therefore almost inseparable from their surrounding buffer zones. When we see the voids of the mini-mountains as more than just residual spaces and instead as settings for spirits, the meaning of the city can suddenly shift. In Seoul, the tangible topography waves can be perceived (in certain locations such as the top of the mini-mountanins) as accentuated by the depth of the sensitive world.

이런 어려움을 겪으면서도 산사는 계속해서 서민들의 지지를 받았다. 불교는 유교와 민속 신앙을 통합하면서 유지되었고, 산사에는 서민들의 발걸음이 끊이지 않았다. 사찰 공동체는 상류층 후원자들에게서 사회적, 정치적 지원과 재정적 후원을 받았다. 새로 생긴 사찰은 일반적으로 규모가 작았지만 불교의 다양한 종파를 수용하면서 여러 개가 세워졌다. 산사는 산비탈 지형에 맞춰 지어졌고, 이런 구조는 한국 고유의 불교 건축물을 탄생시켰으며, 건축 구조는 기하학적인 배치와 달리 지면과 맺는 관계를 통해 유연하게 정의된다.

현재 서울 수도권에는 약 45개의 산사가 있으며, **천장산** 1, **개화산** 24, **봉제산** 19, **국사봉** 12, **서달산** 11 등의 작은 산에 자리 잡은 산사도 있다.

역사 속에서 산사는 격동의 시기를 겪었지만 언제나 자연환경의 보호를 받아왔다. 지역사회 중심에 있는 교회와 달리 산사는 보이지 않는 장애물을 여러 번 넘어야 찾을 수 있다. 산사를 찾는 여정이 산사를 방문하는 것만큼 중요하다고 믿었기 때문에 여정 자체의 본질적인 가치를 강조했다. 산에서는 세속과 분리된다. 산사로 가는 길은 등고선을 따라 초목으로 덮인 장애물을 가로지르며 풍경을 따라 뻗어 있어 거리감과 고립감을 증폭한다. 그래서 산사는 주변을 둘러싼 완충 지역과 떼려야 뗄 수가 없다. 작은 산의 빈 공간을 미개발 공간 이상으로 보고, 나아가 영혼의 터전으로 인식하면 도시의 의미가 아주 달라질 것이다. 서울에서는, 특히 작은 산 정상 같은 특정 장소에서는 실제 지형의 불규칙성이 민감한 현실 세계 속에서 더욱 두드러지게 다가올 수 있다.

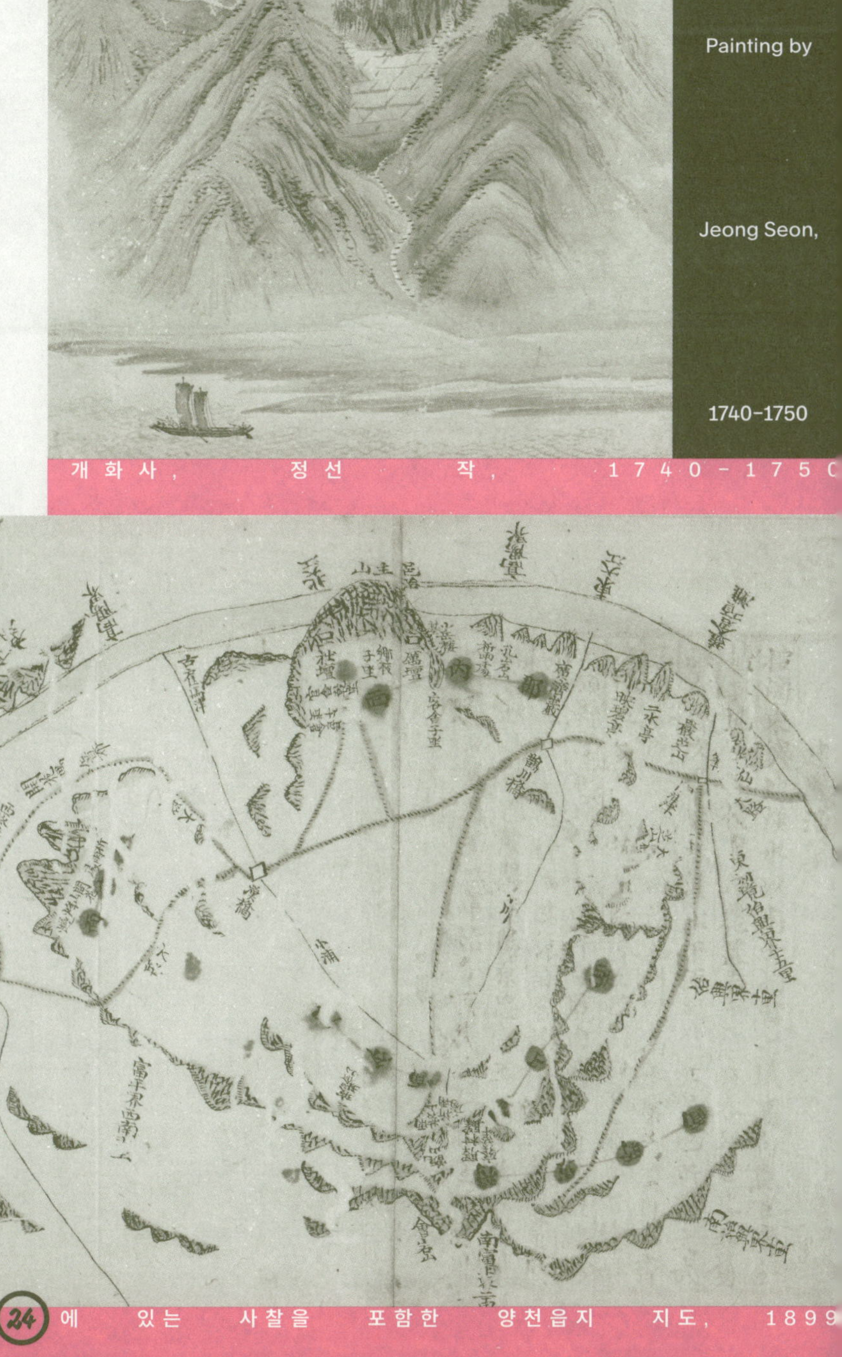

1 Gaehwasa Temple, Painting by Jeong Seon, 1740–1750

개화사, 정선 작, 1740–1750

2 Map in Yangcheon Eupji Including the Temple on 24 개화산 *Gaehwasan*, 1899

개화산 24 에 있는 사찰을 포함한 양천읍지 지도, 1899

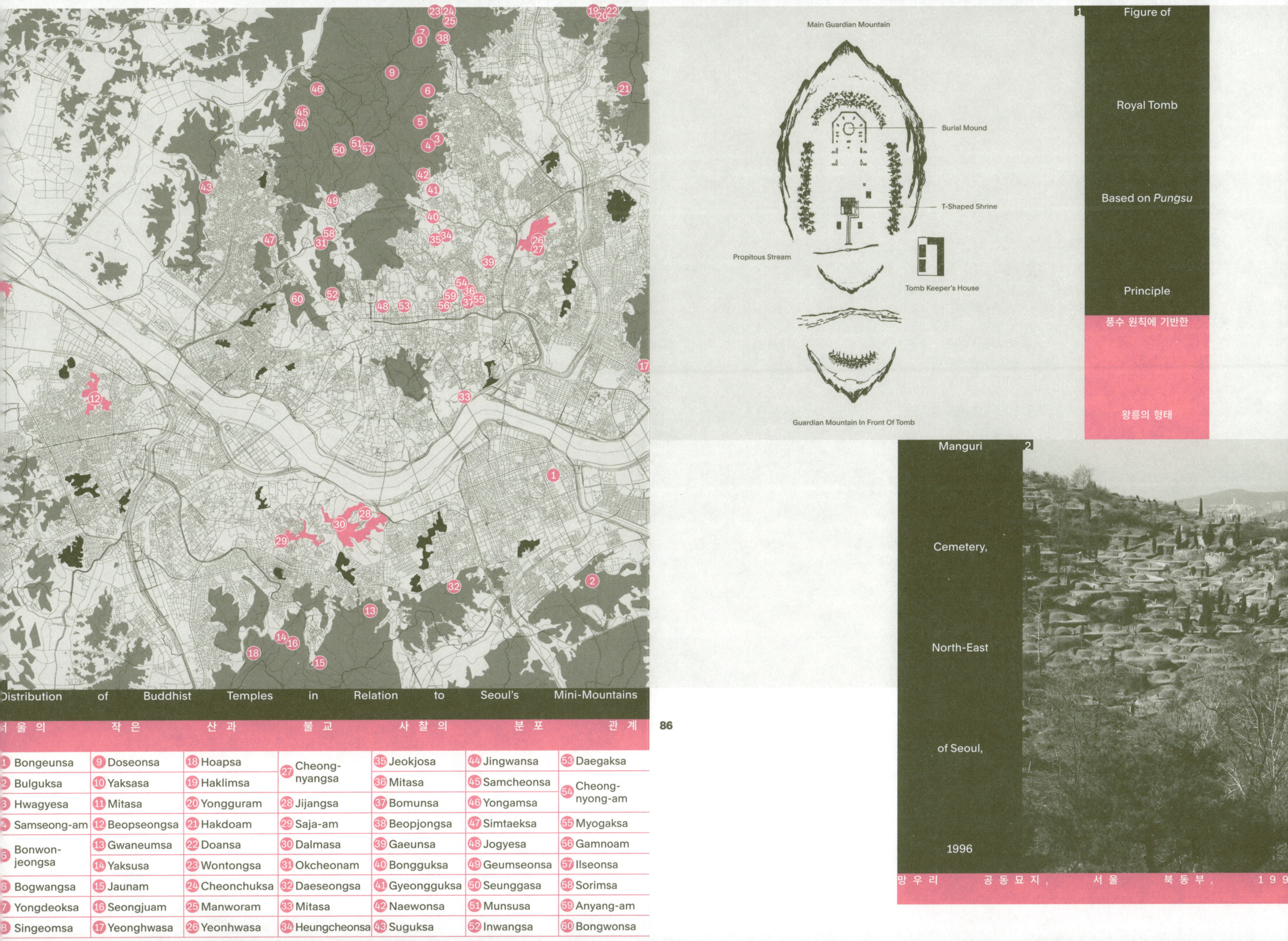

Distribution of Buddhist Temples in Relation to Seoul's Mini-Mountains

서울의 작은 산과 불교 사찰의 분포 관계

1	Bongeunsa
2	Bulguksa
3	Hwagyesa
4	Samseong-am
5	Bonwon-jeongsa
6	Bogwangsa
7	Yongdeoksa
8	Singeomsa
9	Doseonsa
10	Yaksasa
11	Mitasa
12	Beopseongsa
13	Gwaneumsa
14	Yaksusa
15	Jaunam
16	Seongjuam
17	Yeonghwasa
18	Hoapsa
19	Haklimsa
20	Yongguram
21	Hakdoam
22	Doansa
23	Wontongsa
24	Cheonchuksa
25	Manworam
26	Yeonhwasa
27	Cheong-nyangsa
28	Jijangsa
29	Saja-am
30	Dalmasa
31	Okcheonam
32	Daeseongsa
33	Mitasa
34	Heungcheonsa
35	Jeokjosa
36	Mitasa
37	Bomunsa
38	Beopjongsa
39	Gaeunsa
40	Bongguksa
41	Gyeongguksa
42	Naewonsa
43	Suguksa
44	Jingwansa
45	Samcheonsa
46	Yongamsa
47	Simtaeksa
48	Jogyesa
49	Geumseonsa
50	Seunggasa
51	Munsusa
52	Inwangsa
53	Daegaksa
54	Cheong-nyong-am
55	Myogaksa
56	Gamnoam
57	Ilseonsa
58	Sorimsa
59	Anyang-am
60	Bongwonsa

1 Figure of Royal Tomb Based on *Pungsu* Principle

풍수 원칙에 기반한 왕릉의 형태

2 Manguri Cemetery, North-East of Seoul, 1996

망우리 공동묘지, 서울 북동부, 199

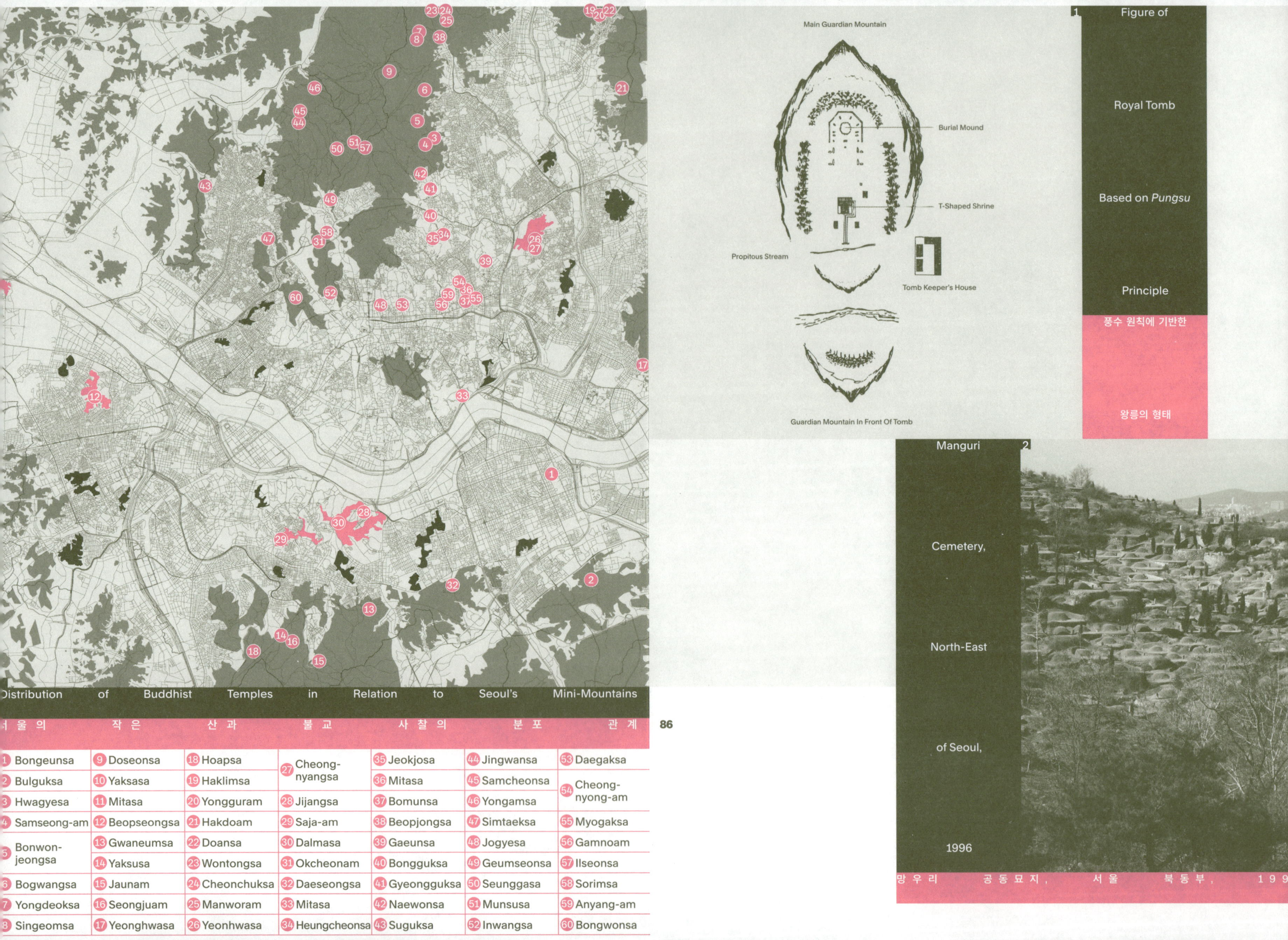

Distribution of Buddhist Temples in Relation to Seoul's Mini-Mountains

서울의 작은 산과 불교 사찰의 분포 관계

1 Bongeunsa	9 Doseonsa	18 Hoapsa	27 Cheong-nyangsa	35 Jeokjosa	44 Jingwansa	53 Daegaksa
2 Bulguksa	10 Yaksasa	19 Haklimsa		36 Mitasa	45 Samcheonsa	54 Cheong-nyong-am
3 Hwagyesa	11 Mitasa	20 Yongguram	28 Jijangsa	37 Bomunsa	46 Yongamsa	
4 Samseong-am	12 Beopseongsa	21 Hakdoam	29 Saja-am	38 Beopjongsa	47 Simtaeksa	55 Myogaksa
5 Bonwon-jeongsa	13 Gwaneumsa	22 Doansa	30 Dalmasa	39 Gaeunsa	48 Jogyesa	56 Gamnoam
	14 Yaksusa	23 Wontongsa	31 Okcheonam	40 Bongguksa	49 Geumseonsa	57 Ilseonsa
6 Bogwangsa	15 Jaunam	24 Cheonchuksa	32 Daeseongsa	41 Gyeongguksa	50 Seunggasa	58 Sorimsa
7 Yongdeoksa	16 Seongjuam	25 Manworam	33 Mitasa	42 Naewonsa	51 Munsusa	59 Anyang-am
8 Singeomsa	17 Yeonghwasa	26 Yeonhwasa	34 Heungcheonsa	43 Suguksa	52 Inwangsa	60 Bongwonsa

1 Figure of Royal Tomb Based on *Pungsu* Principle

풍수 원칙에 기반한 왕릉의 형태

2 Manguri Cemetery, North-East of Seoul, 1996

망우리 공동묘지, 서울 북동부, 199

Tombs

Pungsu is used to organise territory for both the living and the dead, defining the auspicious burial locations for every Korean, from the royal family to common people. Across Korea, Koreans' souls have traditionally been buried under mini-mounds known as tumuli and protected by a man-made semi-circular ridge at the back. This layout reflects the classic *Pungsu* principle: facing south and toward a water stream, with a slope at the back for protection. Each site has an entrance area, a ceremonial area, and the burial mound (*Bongbun*) itself, offering a gradual transition from secular to sacred.

For the royal tombs of the Joseon Dynasty, the most ideal locations were in gentle mountains close to the capital city. As a result, the royal tombs are scattered throughout modern Seoul's central areas. Among the 15 royal tombs located within the Seoul Metropolitan Area, five are located within the city limits and on mountains, including Uireung on (1) 천장산 *Cheonjangsan*, the Tomb of President Rhee Syngman and Seoul National Cemetery on (11) 서달산 *Seodalsan*, and Seonjeongneung and Jeongneung. From above, these areas are easily identifiable as empty meadows of short grass, with artificial topographic feature of mini-mountains, framed by protective forests.

묘지

풍수는 산 사람과 죽은 사람 모두를 위한 구역을 정하는 데 사용되며, 왕실에서 서민에 이르기까지 모든 사람을 위해 길한 묫자리를 알려준다. 한국에서는 사람이 죽으면 전통적으로 인공적인 반원형 등성이 보호하는 봉분 아래 묻혔다. 이런 구조는 전통적인 풍수 원칙에 따라 남쪽을 향하고 앞에는 물이 흐르며 뒤에는 산이 솟아 있어 보호하는 형태다. 각 무덤은 출입구, 제사 장소, 봉분으로 구성되어 세속적인 구역에서 신성한 구역으로 전환된다.

조선시대 왕릉의 가장 이상적인 위치는 수도 근처 완만한 산이었다. 그래서 당시에 세워진 왕릉은 현재 서울 중심지에서 분산되어 있다. 수도권에 위치한 15개의 왕릉 중 5개는 도시 경계와 산중에 자리하는데, **천장산** (1) 에 위치한 의릉, **서달산** (11) 의 이승만 대통령 묘소와 국립현충원, 그리고 선정릉과 정릉이 여기에 포함된다. 이 공간들은 위에서 내려다보면 쉽게 알아볼 수 있다. 짧은 잔디가 펼쳐진 빈 풀밭과 함께 인공적인 지형처럼 솟은 작은 산들이 있고, 그 둘레는 보호림이 감싸고 있다.

1 Tombs on (3) 봉화산 *Bonghwasan*
봉화산 (3) 에 있는 묘소

2

서리풀 (16) 에 있는 묘소

3 Tombs on (16) 서리풀 *Seoripul*

4

개화산 (24) 에 있는 묘소

5 Tombs on (24) 개화산 *Gaehwasan*

f prestigious locations were reserved for royal tombs, larger-cale cemeteries for Seoul's commoners were located on the ides of the capital's hills. In the early Joseon period, burial vas not allowed within Hanyangdoseong and Seongjeosimni. lowever, in the later Joseon period, burial cemeteries became ommon on mountain slopes where no one lived or farmed. 'he living and the dead coexisted harmoniously, with tombs ccupying central places in the city, often in combination with eligious institutions. Between 1912 and 1970, modern cem-etery spaces were established, with public cemeteries main-y located on the city's outskirts. The Japanese "Cemetery Rules" of 1912 banned burial on unrecognised land and egalised cremation. Between 1913 and 1933, 23 public cem-etery sites were designated. However, between the 1930s nd 1970s, these cemeteries were gradually abolished and eplaced with residential areas. During the Korean War, only hose in desperate need of shelter were willing to build homes among the remaining tumuli. In 1973, Manguri Cemetery, the last cemetery on a mountain slope, was closed, signalling a shift toward cremation.

왕릉은 명당 자리를 차지했지만 서울 평민을 위한 대규모 묘지는 수도 언덕 옆쪽에 세워졌다. 조선 초기에는 한양도성과 성저십리 안에 묘지를 만들 수 없었다. 하지만 조선 후기에는 사람이 살지 않거나 농사를 짓지 않는 산비탈에 묘지를 마련하는 일이 흔했다. 때로는 종교적 장소와 결합된 묘지가 도시 중심부를 차지하면서 산 사람과 죽은 사람이 조화롭게 공존했다. 1912–1970년 사이에는 근대적인 묘지 공간이 조성되면서 주로 도시 외곽에 공동묘지가 자리를 잡았다. 1912년 일제강점기의 장묘 제도는, 승인받지 않은 토지에 매장하는 것을 금하고 화장을 합법화했다. 1913–1933년에 23개의 공동묘지가 생겼지만, 1930–1970년대 사이에 공동묘지가 점차 사라지고 주거 지역으로 변모했다. 한국전쟁 중에는 피난처가 절실했던 피난민들이 남아 있던 봉분 위에 집을 지었다. 1973년에는 산비탈을 차지했던 마지막 공동묘지, 망우리 묘지가 폐쇄되면서 장례 풍습이 화장으로 점점 전환되었다.

1 Temple on 12 국사봉 Guksabong

국사봉 12 에 있는 사찰

2 Temple on 24 개화산 Gaehwasan

개화산 24 에 있는 사찰

Today, some tumuli still punctuate the mini-mountains. In (24) 개화산 *Gaehwasan*, the tombs of the Sim family, who held different government positions in the early and mid Joseon dynasty, which were also recorded in Yangcheon Eupji as the 'Pungsan Sim Clan Munjeonggong Cemetery' occupy the southern slopes of the mountain, overlooking Gimpo Airport. In (3) 봉화산 *Bongkwasan*, some tumuli still coexist among fruit trees and community gardens, hinting at the lived history of the area.

지금도 작은 산에는 봉분들이 남아 있다. **개화산**(24)에는 조선 초기와 중기에 여러 벼슬을 지낸 심씨 가문 묘역이 김포공항이 내려다보이는 산 남쪽 경사면에 있으며, 이 묘역은 『양천읍지(陽川邑誌)』에도 "풍산 심씨 문정공파 묘역"으로 기록되어 있다. 과수원과 마을 공원 사이에 남아 있는 **봉화산**(3)의 봉분 몇 개는 지역의 오랜 역사를 암시한다.

Chapter 2

2 장

Boundaries
경계

The Voids' Contours
보이드의 윤곽

서울의 지형은 물리적 제약을 주므로 도시를 개발할 때 가장 먼저 마주하는 한계로 이해된다. 빈 공간으로 남겨진 산 정상을 둘러싼 건물들, 그리고 미개발 마을에서 아파트 단지, 시민 공원으로 변모한 도시 구조를 살펴보면 이 한계를 만들어낸 사회적, 정치적 힘이 드러난다. 지형, 솔리드와 보이드, 접근성, 작은 산을 둘러싼 지적선 등 서로 다른 경계선이 어긋나는 지점을 들여다보면 그 경계들에 대한 다양한 해석이 가능해진다. 나아가 이 지점들은 우리가 도시를 인식하고 표현하는 방식에 의문을 던진다.

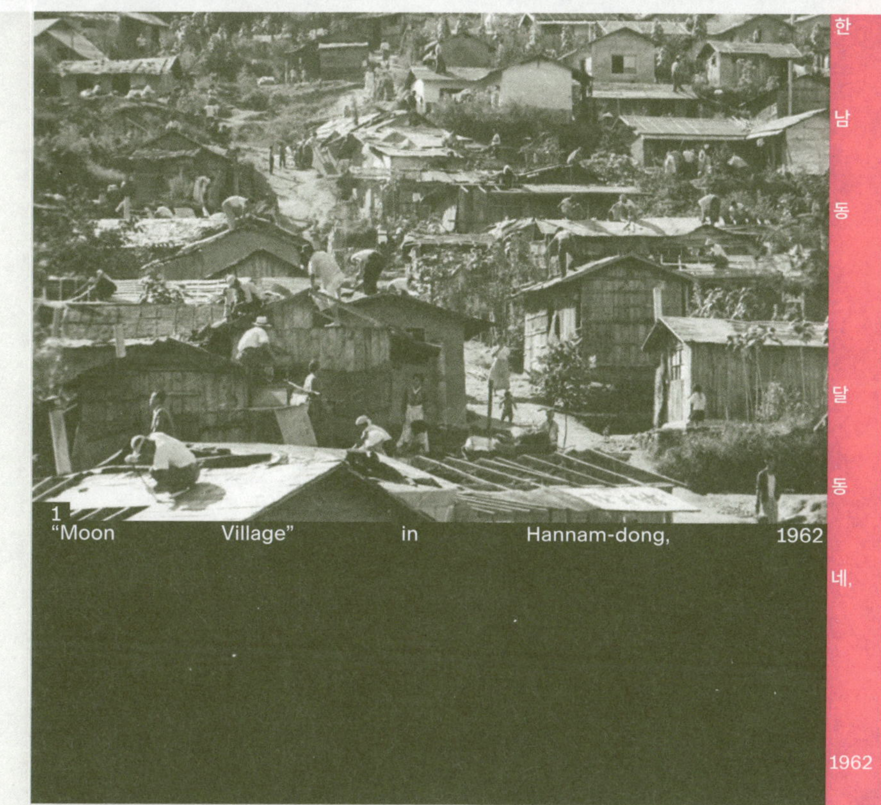

1
“Moon Village” in Hannam-dong, 1962

Because of the physical constraints it represents, the topography of Seoul can be seen as the first tangible limit on built development within the city. Examining the belt of buildings around the unbuilt summit of the hills, and the transformation of the urban texture (from squatter villages to apartment blocks to citizen parks) indicates the underlying social and political forces shaping this limit. Revealing the mismatching of different boundary lines—whether through topography, solid and void, accessibility, or cadastral lines surrounding the mini-mountains—brings forward multiple interpretations of these thresholds, leading us to question our representation of the city.

Because of the physical constraints it represents, the topography of Seoul can be seen as the first tangible limit on built development within the city. Examining the belt of buildings around the unbuilt summits of the hills, and the transformation of the urban texture (from squatter villages to apartment blocks to citizen parks) indicates the underlying social and political forces shaping this limit. Revealing the mismatching of different boundary lines—whether through topography, solid and void, accessibility, or cadastral lines surrounding the mini mountains—brings forward multiple interpretations of these thresholds, leading us to questioning our representation of the city.

Process of Urbanisation

The accelerated process of urbanisation in Seoul began n the 1960s, after the Japanese Colonial Period (1910–1945) nd the Korean War (1950–1953). Attracted by the development of labour-intensive industries, a massive influx of rural nigrants moved to Seoul, resulting in a rapid increase in population, which soared from 1.6 million in 1955 to 2.4 millions n 1960 and 10.6 million by 1990. (Up until the mid 1980s, vhen Seoul's industrialization reached its peak, the population ncreased at an average of nearly 300,000 people per year). s the network of infrastructure expanded, the city's footprint xtended in all directions, quickly occupying flat and easily ccessible land. However, despite their strategic location vithin the heart of the growing city, the steep slopes of hills lid not follow the same rapid pace of urbanisation. The difficult ccessibility, rocky terrain, and lack of infrastructure made hese areas less desirable for development. The less favourable sites within the Seoul Metropolitan Area became a refuge for those less favoured, including both the new migrants and those displaced from the city centre. The city's poorest residents adapted to these constraints, using their ingenuity to establish their homes on the slopes, creating what is known as "moon villages."

도시화 과정

도시화는 일제강점기(1910–1945)와 한국전쟁(1950–1953)이 지나고 1960년대에 들어서 가속화됐다. 노동집약적산업 발달로 농촌인구가 서울로 대규모 이주했고, 그 결과 서울 인구는 1955년 160만 명에서 1960년 240만 명, 1990년에는 1,060만 명으로 급증했다(서울의 산업화가 절정에 달했던 1980년대 중반까지 연평균 인구 증가 수는 30만 명이었다). 사회 기반 시설의 연결망이 구축되고 도시 면적이 사방으로 확장하면서 평평하고 접근하기 쉬운 토지를 빠르게 잠식해 나갔다. 하지만 이렇게 성장하는 도시의 중심부라는 전략적 위치가 무색하게, 여러 산의 가파른 경사면은 도시화의 빠른 속도를 따라가지 못했다. 떨어지는 접근성, 암산 지형, 사회 기반 시설의 부족 때문에 산비탈은 개발 지역으로 선호되지 않았다. 수도권의 이런 지역은 도시로 새로 이주한 사람과 중심지에서 쫓겨난 사람을 포함한 소외 계층의 피난처가 되었다. 도시 최빈층은 산비탈에 독창적으로 집을 지어 '달동네'를 형성하면서 지형의 한계에 적응했다.

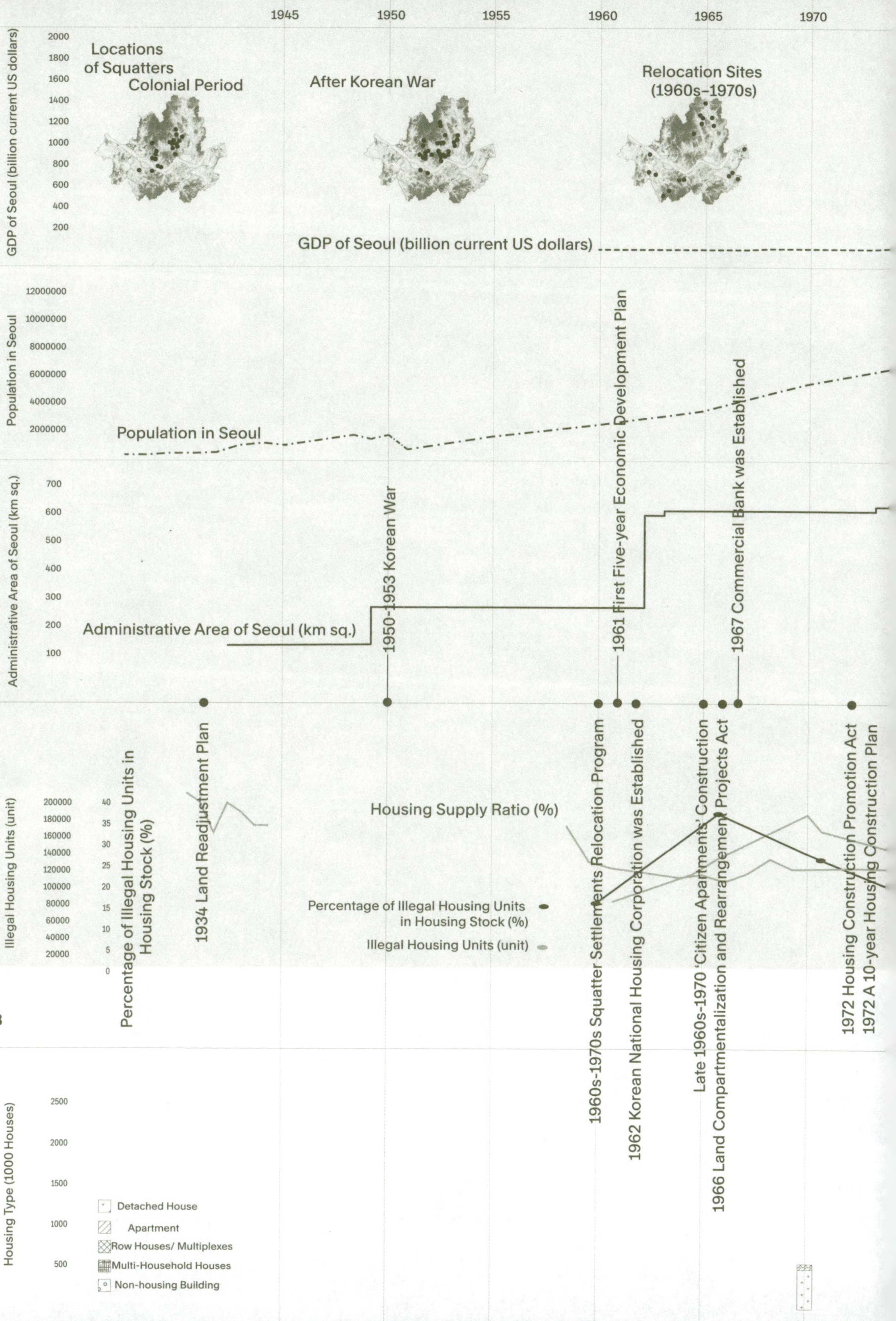

1945 1950 1955 1960 1965 1970

GDP of Seoul (billion current US dollars)
2000 1800 1600 1400 1200 1000 800 600 400 200

Locations of Squatters
Colonial Period
After Korean War
Relocation Sites (1960s–1970s)

GDP of Seoul (billion current US dollars)

Population in Seoul
12000000 10000000 8000000 6000000 4000000 2000000

Population in Seoul

Administrative Area of Seoul (km sq.)
700 600 500 400 300 200 100

Administrative Area of Seoul (km sq.)

1950-1953 Korean War
1961 First Five-year Economic Development Plan
1967 Commercial Bank was Established

Illegal Housing Units (unit)
200000 180000 160000 140000 120000 100000 80000 60000 40000 20000

40 35 30 25 20 15 10 5 0

Percentage of Illegal Housing Units in Housing Stock (%)

1934 Land Readjustment Plan

Housing Supply Ratio (%)

Percentage of Illegal Housing Units in Housing Stock (%)
Illegal Housing Units (unit)

1960s-1970s Squatter Settlements Relocation Program
1962 Korean National Housing Corporation was Established
Late 1960s-1970 'Citizen Apartments' Construction
1966 Land Compartmentalization and Rearrangement Projects Act
1972 Housing Construction Promotion Act
1972 A 10-year Housing Construction Plan

Housing Type (1000 Houses)
2500 2000 1500 1000 500

Detached House
Apartment
Row Houses/ Multiplexes
Multi-Household Houses
Non-housing Building

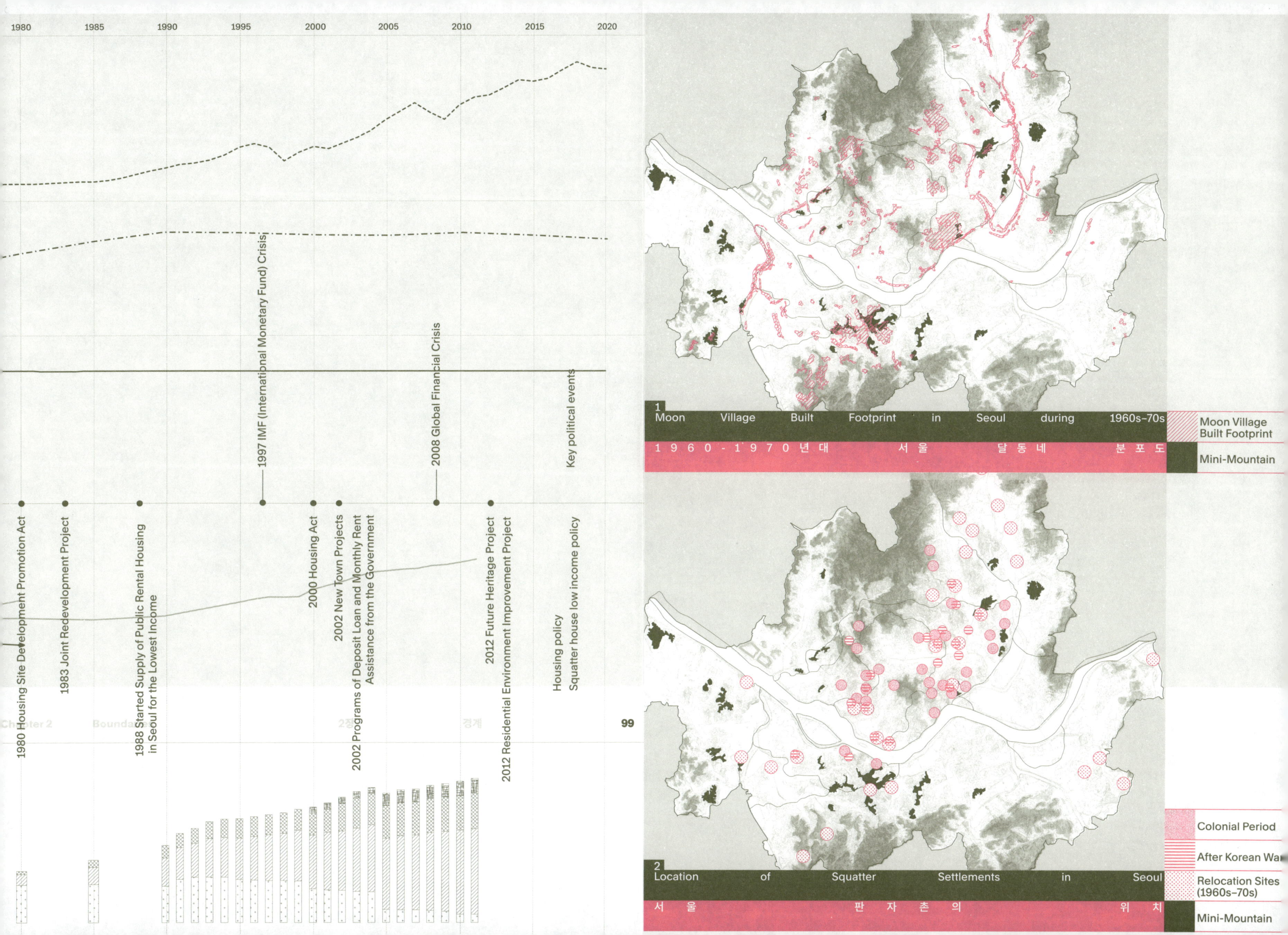

1 Moon Village Built Footprint in Seoul during 1960s–70s

1960-1970년대 서울 달동네 분포도

2 Location of Squatter Settlements in Seoul

서울 판자촌의 위치

Moon Villages (Living on the Slopes)

The term "moon villages" or "Dal-dong-ne" (달동네) is a oetic name for the squatter settlements that mainly occupied ıe less favourable hillsides of the city. For some, the name epresents closeness to the sky, because most of these settle-ıents are located at the top of hills. For others, it describes ow the moonlight shone through the poor-quality materials f temporary shelters in hillside settlements.These habitations ere not only characterised by their cheap materials but also uffered from a lack of infrastructure.

During the Japanese Colonial Period and after the orean War, the squatter settlements were mainly located round the inner mountains surrounding the old Seoul City. owever, due to urbanisation, these sites needed to be leared for development. The government, therefore, con-ucted a relocation programme, removing squatters from the ity centre and relocating them to around 20 sites within the developing Seoul Metropolitan Area. With the continuous flux of migrants from the countryside, the number of units in infor-mal settlements actually increased during the 1960s and 1970s. According to *Seoul Statistical Yearbook* The number of shan-ties units (including board shacks, tents and mud huts) grew from 41,238 in 1961 to 187,500 units (including board shacks and illegal structures) in 1970 or 30 percent of the total housing units in Seoul.

달동네(산비탈 거주지)

달동네는 도시에서 누구도 선호하지 않던 언덕을 차지한 판자촌을 낭만적으로 부르는 명칭이다. 언덕 위에 있어 하늘에 떠 있는 달과 가깝다는 데서 유래했다고 해석하기도, 비탈길에 위치한 임시 주택의 부실한 건축자재 사이로 스며드는 달빛을 묘사한 이름이라고 해석하기도 한다. 달동네 주거지는 부실한 자재로 지어졌을 뿐 아니라 기반 시설이 부족해 살기에 열악한 환경이었다.

일제강점기와 한국전쟁 이후 판자촌은 주로 구시가지를 둘러싼 산 주변에 자리 잡았다. 하지만 도시화가 진행됨에 따라 개발을 위해 판자촌을 철거해야 했다. 정부는 먼저 개발에 착수한 수도권 내 20여 개 지역으로 판자촌 주민을 이주시키는 사업을 시작했다. 시골에서 계속 이주해 오는 사람들로 1960년대와 1970년대 판자촌 주택 수는 실제로 증가했다. 『서울통계연보』에 따르면 판자촌 주택(판잣집, 텐트, 진흙집 포함) 수는 1961년 4만 1,238가구에서 1970년 18만 7,500가구(판잣집, 무허가 건축물 포함)로 증가했고, 이 수치는 서울 전체 주택의 30퍼센트를 차지했다.

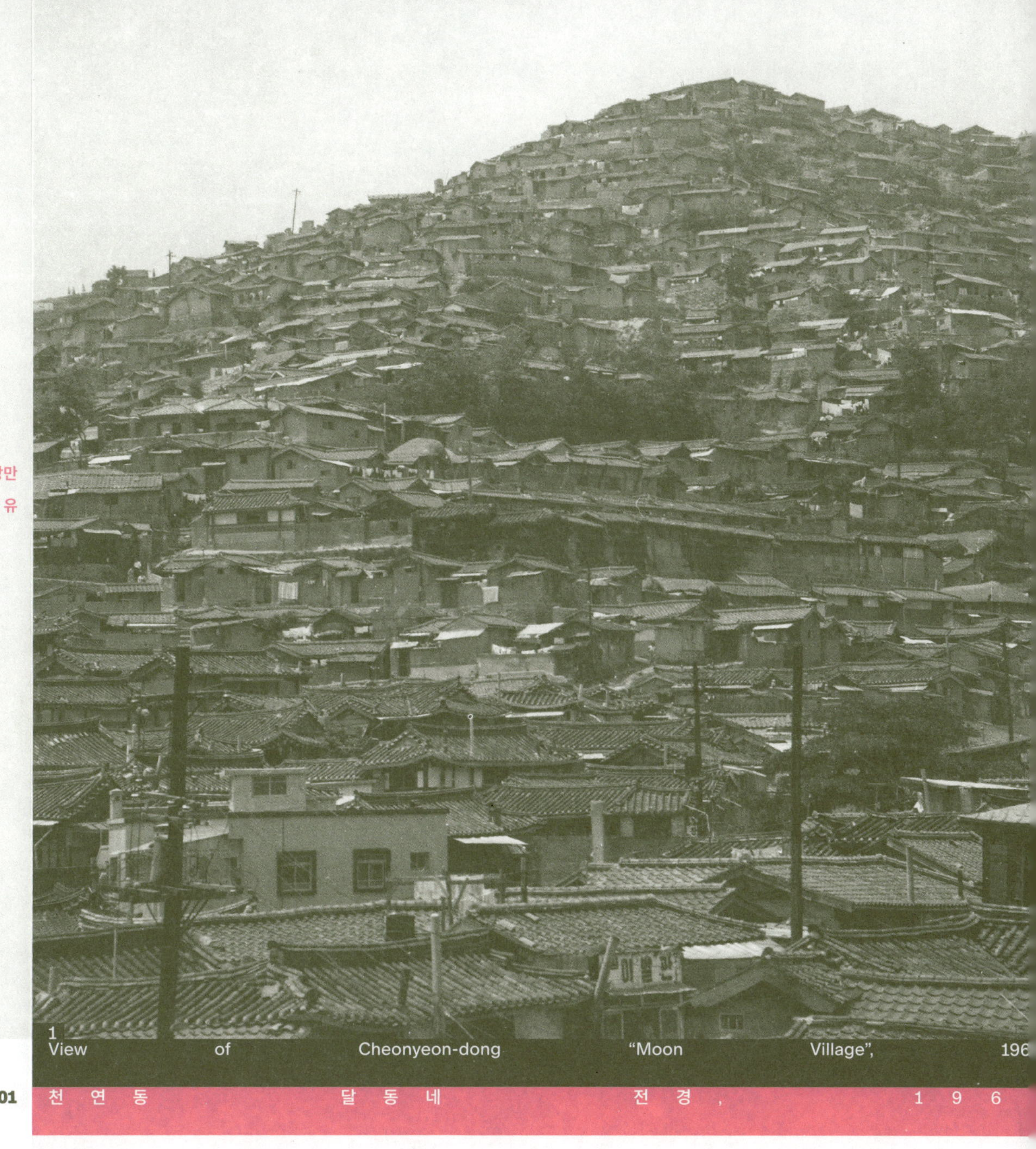

1 View of Cheonyeon-dong "Moon Village", 196
천연동 달동네 전경, 196

Although the sites were allocated by the government, most residents could not afford to pay for property rights. Small parcels of approximately 26–40 square metres were distributed by the government to squatter families, with each provided with one tent and 200 mud bricks to build their houses. These limited resources often resulted in poor-quality housing. Thanks to the construction skills of their owners and their understanding of the terrain, however, some units became models of pragmatism and ingenuity, with sheds and access paths following the natural contours of the land.

The urgent need for housing is reflected in the more opportunistic layout of these settlements. Inhabitants had to rely on their skills to adapt to the slopes. Unlike traditional Korean villages, these settlements could not be built according to the principles of pungsu, which favour placing houses on a north-south axis. The sites allocated by the government were usually located on the highest parts of hills, with insufficient infrastructure and unfavourable orientations for sunlight and ventilation.

정부에서 판자촌 주민의 이주지를 할당했지만 대부분은 이주권을 구입할 형편이 되지 않았다. 정부는 판자촌 주민들이 직접 집을 짓도록 각 가구에 26–40제곱미터의 작은 공간과 천막 하나, 진흙 벽돌 200개를 제공했다. 이런 자재로 지은 주택의 질은 물론 형편없었지만, 주택 소유자들은 지형을 먼저 이해한 후 건설 기술을 활용해 지형의 원래 윤곽을 따라 창고와 출입문 등을 만들었다. 이런 특징을 지닌 일부 주택은 실용적이고 독창적인 주택 모델이 되었다.

주민들에게 주택이 얼마나 절실했는지는 달동네의 임시방편적 주거 구조에서도 드러난다. 비탈길에 집을 지을 때는 자기만의 기술에 의존해야 했다. 그래서 달동네 집들은 전통적인 마을처럼 풍수 원칙에 따라 남북 축선에 맞춰 배치할 수 없었다. 정부가 할당한 부지는 대부분 언덕의 가장 높은 지역이어서 기반 시설이 부족했고, 해도 잘 들지 않고 환기도 잘 되지 않았다.

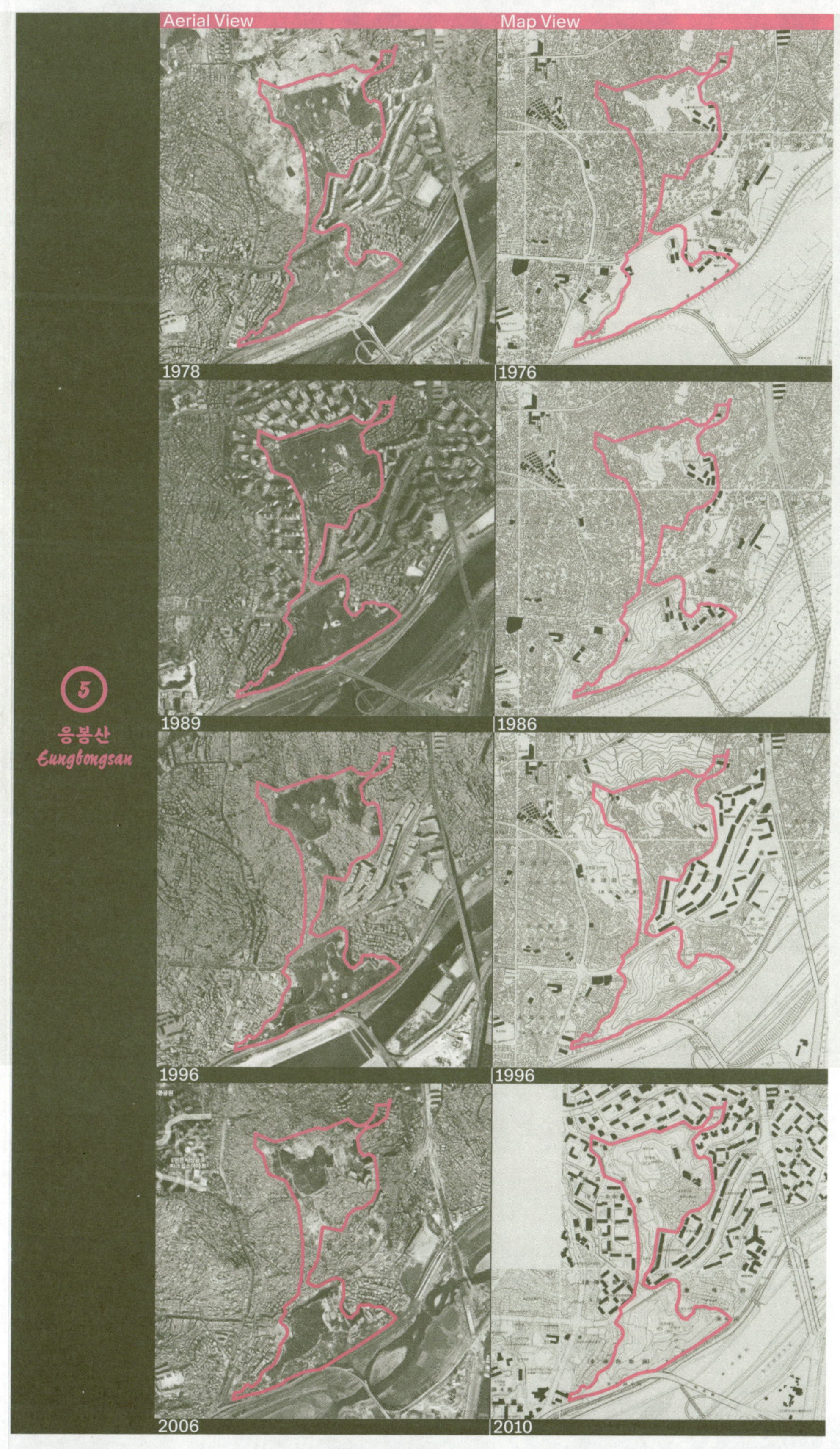

is within these meanders that a large part of Seoul's current opulation first saw the sky. According to statistics, 20–30% of ne urban population once resided in hillside squatter settle- nents, and Seoul's "panjachon" or literally wood board village, veraged 7.1 residents per unit. Despite the difficult conditions, ome still remember with nostalgia the sense of a village life- tyle within the heart of the new city. Even amid the hardships, hese hillside homes often caught a refreshing breeze, and ome had spectacular views. Yet, in Seoul, unlike other cities, vhere views typically drive up property prices, the stigma of overty attached to these hillside squatter settlements remains.

Mapping records of Seoul's shanty towns during the 960s–70s show that the mini-mountain of 5 응봉산 *Eungbongsan* was fully occupied by one of the largest "moon illages" at that time. Going back to the 1946 map (Chapter 1), ve can hardly see any building occupation of the area, except erhaps the presence of a cemetery on the north-western lope (indicated on the 1922 map). The woodland covered the northern art of the mountain and the lower urrounding area of the southern nountain, whereas the surrounding latter ground was covered by grass- and. Due to the steepness of the cliff, vhich is composed of hard granite ock, the southern part of the moun- ain mostly remained undeveloped hroughout the whole history of the city, even though some archive pictures show the presence of a few houses cling- ing to the cliff. In contrast, the northern area became, in less than 20 years, the way it is shown on the 1966 map (Chapter 1), fully covered by the "moon villages". More precise records that identify the size, shape, and location of the housing units reveal that in 1976, only the very top of the hill was unbuilt. The occupation of the hill and its surroundings covered an area similar to the footprint of Namsan, making the 5 응봉산 *Eungbongsan* region the biggest informal settlement in the city.

현재 서울 인구 중 상당수가 이런 구불구불한 골목의 달동네에서 처음으로 하늘을 보았다. 통계에 따르면 도시 인구 중 20–30퍼센트가 한때 산비탈의 판자촌에 거주한 경험이 있고, 당시 서울 판자촌 주택의 평균 가구원 수는 평균 7.1명이었다. 거주환경이 이렇게 열악했음에도 새로운 도시의 중심부에 살았던 때를 그리워하는 사람도 있다. 힘든 생활이었지만 산비탈에 위치한 집은 시원한 바람이 통했고 멋진 전망을 즐길 수도 있었다. 하지만 전망에 따라 부동산 가격이 오르는 다른 도시와 달리, 언덕 위 서울 판자촌에 대한 빈곤의 낙인은 여전히 남아 있다.

1960–1970년대 서울 판자촌 지도를 보면 당시 최대 규모의 달동네가 **응봉산** 5 을 점유했음을 알 수 있다. 1946년 지도(1장)에서는 북서쪽 경사면에 묘지(1922년 지도에 표시)를 제외하고는 건물을 거의 찾아볼 수 없다. 응봉산 북쪽과 남쪽 낮은 지대는 삼림으로 덮여 있고 주변 평지는 초원이었다. 단단한 화강암 절벽의 가파른 경사 때문에 응봉산 남쪽은 도시화 과정에서 대부분 미개발 지역으로 남았지만 일부 사진 기록물에서는 절벽에 매달리다시피 세워진 집 몇 채를 확인할 수 있다. 이와 대조적으로 응봉산 북쪽은 20년도 채 되지 않아 1996년 지도(1장)에 보이는 것처럼 달동네로 뒤덮였다. 주거지의 크기, 형태, 위치 등을 더 자세히 기록한 자료에 따르면 1976년에는 응봉산 정상만이 빈 공간이었다. 당시 응봉산과 인근 판자촌이 차지한 면적은 남산과 비슷할 정도였으므로 **응봉산** 5 지역은 서울에서 가장 큰 규모의 판자촌 정착지였다.

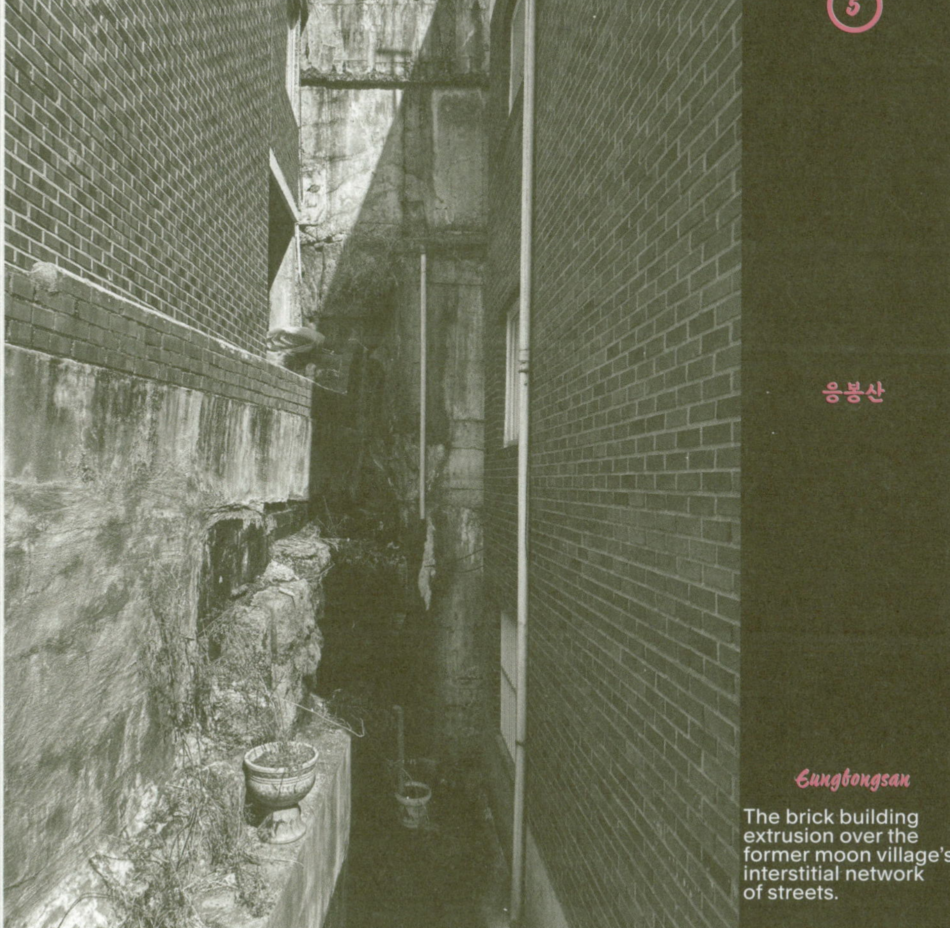

1 5

응봉산

Eungbongsan

The brick building extrusion over the former moon village's interstitial network of streets.

2 3

응 봉 산 5

이전 달동네의 복잡한 골목길 위로 솟은 벽돌 건물

In the late 1960s, aware of the harsh living conditions within the informal settlement, the government initiated the Residential Environment Redevelopment Programme, with the mission of relocating "moon villagers" into standardised apartment blocks. From the 1980s, the northern and western slopes first undertook a clearing of the "moon village" occupation, creating a surrounding buffer for a while. But this clearing movement spread to the top of the hill, and as the 2010s map shows, the buffer zone was converted into a neigborhood of apartment blocks.

A similar transformation applied to most of the "moon villages" in the city at that time. Still, it is worth noting the particular case of Eungbong-dong 265, a living area on the mountain that has survived until today. While the other living areas were demolished and relocated to apartment blocks at the foot of the hill, the low-rise neighbourhood redeveloped itself through the "local improvement method". This evolution of government policies reflects the wish to reduce the negative impact of relocation on the lives of the residents. Concrete and brick buildings of 5 storeys replaced the "moon village" units, but the intricate street networks typical of moon villages remained. The extrusion of these blocks over the labyrinthine division of the ground gives the impression of a fortress strongly anchored and protecting the souls of the neighbourhoods. Here and there, between the interstitial gaps, emerge a micro-garden, and some kimchi pots—indicators of what life on this unfavourable ground might once have been like.

1960년대 후반, 판자촌의 혹독한 생활환경을 인지한 정부는 달동네 주민을 표준화된 아파트 단지로 이주시키는 계획과 더불어 주거 환경 개선 사업을 시작했다. 1980년대부터 북쪽과 서쪽 산비탈의 달동네를 철거하면서 잠시 동안 주변에 완충 지대가 조성되었다. 하지만 이런 철거 사업은 나중에 산 정상까지 영역을 확장했고, 2010년대 지도에서 볼 수 있듯이 완충 지역은 아파트 단지로 교체됐다.

다른 달동네에도 비슷한 변화가 있었다. 하지만 지금껏 살아남은 산 근처 달동네인 응봉동 265번지의 특별한 사례는 주목할 만하다. 다른 달동네는 철거돼 산기슭의 아파트 단지로 이전된 반면, 낮은 지대의 달동네는 '현지 개량법'을 통해 자체적으로 재개발되었다. 이는 이주 과정에서 주민들이 겪을 수 있는 부정적 영향을 줄이려는 정부 정책의 변화에서 비롯된 것이었다. 5층 규모의 콘크리트 건물과 벽돌 건물이 달동네를 대체했지만, 달동네 특유의 구불구불한 길은 여전히 남아 있다. 미로처럼 구획된 땅 위로 솟은 건물 블록들은 마치 이웃을 보호하는 견고한 요새처럼 보인다. 여기저기 건물 사이로 아주 작은 마당과 김칫독이 드러나, 이 척박한 땅 위에서의 삶이 한때 어떠했는지를 짐작하게 한다.

1 Construction of the Geumhwa Citizen Apartment, Yongsan District, 1968
용산구 금화시민아파트 건설 현장, 1968

2

3

4 Geumhwa Citizen Apartment, 1971
금화시민아파트, 1971

Apartment Blocks (The Collapse Incident)

Within 40 years of turmoil, the framing boundaries of the unbuilt mountain's summit radically changed texture from a light carpet of fragile shelters to robust and standardised 20–30 storey apartment blocks.

From the late 1960s the city government aimed to improve conditions for residents while addressing the perceived chaos in certain areas of the city. This initiative meant that the moon villages started to be under threat. Launched by the Seoul Metropolitan Government, housing policy including the Housing Construction Promotion Act and Urban Development Act intensified large-scale development, which promoted the construction of standardized apartment blocks. This policy led to a decline in the number of moon villages. The shift from small one-story squatter houses to large apartment blocks changed Seoul's housing landscape. A report on housing types in Seoul showed that in 1970, 88% of homes were detached houses and only 4% were apartments. However, by the year 2000, the balance had changed, with 51% of housing being apartments and only 26% being detached houses. To make way for these new developments, the government cleared dense areas that covered the hills almost to the top, focusing construction on the lower part of the slopes.

아파트 단지(붕괴 사고)

40년 간의 혼란 속에서, 산 정상부 주변의 빈 공간은 금방이라도 무너질 듯한 판자촌에서 표준화된 20-30층짜리 아파트 단지로 근본적인 변화를 겪는다.

1960년대 후반부터 서울시는 특정 지역의 혼란을 해소하고 주민들의 생활환경을 개선하는 것을 정책 목표로 삼았다. 이 사업으로 달동네는 본격적으로 위협받기 시작했다. 서울시가 추진한 주택건설촉진법과 도시개발법 등의 주택정책으로 대규모 개발이 진행되면서 표준화된 아파트 단지가 조성됐고, 이로 인해 달동네 수는 크게 줄어들었다. 작은 1층짜리 판잣집들이 대형 아파트 단지로 바뀌면서 서울의 주거 풍경이 변화했다. 1970년 서울 주택 유형을 보고한 자료에 따르면 당시 주택의 88퍼센트가 단독주택이었고 아파트는 4퍼센트에 불과했다. 그러나 2000년에는 아파트가 전체 주택의 51퍼센트를 차지하고 단독주택은 26퍼센트로 줄어들면서 상황이 역전됐다. 새로운 개발 지역을 확보하기 위해 정부는 산 정상을 덮고 있던 가건물을 철거하고 산비탈의 저지대에 건설을 집중했다.

1

2
The Collapse of Wau Citizen Apartment on 6 와우산 *Wausan*, April 8, 1970

와우산 6 의 와우시민아파트 붕괴, 1970.4.8

As the Seoul Metropolitan Government embarked on a redevelopment project to clear unauthorised shantytowns scattered throughout the city, they built what they called "Citizen Apartments" to house those displaced. The government attempted to relocate squatter residents into "citizen apartments" constructed near their original areas. However, these apartments were designed for middle-class residents, leading to skyrocketing property values. As a result, only about 20% of original residents could afford to move back to the apartments in their old neighbourhoods. Additionally, relocation sites were in short supply, leading to the emergence of a new type of illegal settlement called "vinyl house villages" (plastic house villages) from the mid-1980s. The "moon villages" that survived under JRP often did so because of physical constraints or low desirability of their locations.

The "citizen apartments" were constructed with public funding providing the structural skeleton, while the residents were responsible for building the rest, including interior spaces and exterior walls. The plan was to construct 2,000 buildings, providing 90,000 households, over three years, starting in 1969. From 1969, 32 shantytowns were demolished and the first year, 406 buildings were constructed, accommodating 15,840 households. In 1970, the poor quality of these apartments was to lead to disaster.

On April 8, 1970, Wau Citizen Apartment, located on the northern part of 6 와우산 *Wausan*, slid down the hill, resulting in 33 deaths and 40 injuries. The collapse occurred just four months after the building's completion ceremony, which had been attended by President Park Chung-hee, also known as "President Bulldozer."

서울시는 도시 전역에 분산된 무허가 판자촌을 철거하는 재개발 사업을 시작하며 주거지를 잃은 주민을 위해 '시민아파트'를 건설했다. 정부는 거주지 근처에 준공된 시민아파트로 판자촌 주민을 이주시킬 계획이었지만, 시민아파트는 실상 중산층을 위한 주거지였기 때문에 가격이 급등했다. 그 결과, 원래 판자촌에 거주하던 주민 중 약 20퍼센트만이 인근 시민아파트의 입주권을 살 수 있었다. 한편, 이주할 수 있는 부지가 부족했기 때문에 1980년대 중반부터는 '비닐하우스촌'이라고 불리는 새로운 형태의 무허가 정착지가 등장하기도 했다. 도시재개발 사업에서 살아남은 일부 달동네는 물리적 제약이나 낮은 입지 선호도 덕분에 존속할 수 있었다.

시민아파트는 공공 예산으로 건물의 기본 골조만 완공하고, 실내 공간과 외벽을 포함한 나머지 구조는 입주자들이 책임져야 했다. 1969년부터 3년 동안 총 2,000동의 시민아파트를 건축해 9만 가구를 공급할 계획이 세워졌다. 이에 따라 1969년에는 32개 판자촌을 철거하고, 1년 동안 시민아파트 406동을 지어 15,840가구를 수용할 수 있도록 했다. 그러나 1970년, 시민아파트의 부실 공사가 대형 재난으로 이어졌다.

같은 해 4월 8일, 6 **와우산** 북쪽에 위치한 와우시민아파트가 산비탈을 따라 무너지면서 33명이 사망하고 40명이 부상을 입는 사건이 발생했다. '불도저 대통령'으로 불리던 박정희 대통령이 준공식에 참석한 지 불과 4개월 만에 발생한 붕괴 사고였다.

2 Seoul Panorama View from Ansan Looking at Namsan, 1920/2009

안산에서 바라본 서울 남산 전경, 1920/200

he investigation into the disaster revealed severe flaws in ne construction of Citizen Apartments. It turned out that the udget invested in constructing an apartment complex on a teep mountainside was less than half of the average price nd this was reflected in the drastic decline in quality of con-truction material (in particular in particular the poor quality f the concrete and the reduced number of structural steel ars which forms the skeleton of the building) and that the onstruction timeline (only six months) was unrealistic. Fol-owing the Wau Citizen Apartment collapse, an inspection f the entire Citizen Apartment scheme was carried out cross Seoul. During the 1970s alone, about 100 Citizen partments were demolished as a result of the findings from nese inspections.

The collapse of the Wau Citizen Apartment, which was sign of corruption and poor construction practices, affect-d perceptions of living on the slopes. This led to new efforts to reinforcing the walls and rigidity of the mountain's slope, but at the same time, it increased awareness of natural assets. Moving away from planning only for the built environment, the city launched green spaces policies and park regulations to preserve natural spaces in the city. As part of the initiative, in 1972 the government implemented regulations on building height and limited high-rise in designated areas around the major mountains as part of efforts to protect the urban landscape and prevent over-crowding. As a result of all these efforts to improve the living environment through-out Seoul, the materiality of the figure-ground boundary on the mini-mountains deeply transformed with a clearer partition between the solid (living spaces) and the voids (areas designated for recre-ational activities and wander-ing).

조사 결과, 부실 공사가 붕괴 원인으로 드러났다. 가파른 산비탈에 아파트 단지를 건설하기 위해 투입된 예산은 평균의 절반에도 미치지 못했고, 이는 건설 자재의 품질 저하(특히 저품질 콘크리트와 건물 골조에 쓰이는 철근 부족)와 비현실적으로 짧은 공사 일정(고작 6개월)으로 이어졌다. 와우시민아파트 붕괴 사고 이후 서울 전역의 시민아파트에 대한 전수 점검이 이루어졌고, 1970년대에만 약 100동의 아파트가 철거되었다.

부패와 부실 공사의 산물인 와우시민아파트 붕괴 사고는 산비탈 거주지에 대한 인식을 전환시키는 계기가 되었다. 이후 산 경사면에 세워진 건축물의 강도와 벽을 보강하려는 노력이 시작되었고, 도시 내 자연환경에 대한 인식도 높아졌다. 서울시는 건축 환경만을 강조했던 기존 계획에서 벗어나 도시 속 자연 공간을 보존하는 녹지 조성 정책과 공원 규정을 마련했다. 이 계획의 일환으로 1972년 정부는 도시경관을 보호하고 과밀 현상을 억제하기 위해 주요 산 주변의 건축물 높이를 제한하고 고층 건물 건설을 규제하는 정책을 시행했다. 서울 전역의 생활 환경을 개선하려는 노력 덕분에 작은 산들 주변의 건물-지대 형태는 솔리드와 보이드의 구분이 더욱 확실해지는 형태로 변모했다.

1 Popular dust & tick hoovers, are to be found at the bottom of the urban hills, allowing a comfortable return to the city after a quick break in the forest.

산자락에 설치된 먼지 및 진드기 제거기는 사람들이 애용하는 시설로, 숲에서 잠시 휴식을 취한 후 편안히 도시로 돌아가게 해준다.

Accessibility (The Hard & Soft Boundaries)

Shaped by both natural topography and human intervention, the boundary line between the urban fabric and the natural places has undergone successive fluctuations. At each mini-mountain, the type of boundary varies according to its physical presence and the transversive possibilities it offers. From hard barriers like concrete retaining walls and buildings, to softer transitions like parks and green spaces, these boundaries influence how people interact with the environment and perceive the space around them.

In certain instances, such as at (6) 와우산 *Wausan*, retaining walls are used to level slopes into terraces for apartment blocks, reinforcing the solid and impassable edge of the mountain. This influences the way visitors navigate the area, leading to dead ends and blocked routes until they discover access through the park (the famous slope that was converted into a park after the apartment block collapse).

At (20) 우장산 *Ujangsan*, pedestrian paths cross through a series of parks, allowing for a more fluid transition into the hills. However, in some cases, the perception of boundaries can be more ambiguous. At (1) 천장산 *Cheonjangsan*, for example, the presence of sparsely built university campuses and other institutions surrounding the foot of the mountain offers an illusion of porosity, but these open spaces provide few actual access points to the slopes.

접근성(단단한 경계와 유연한 경계)

인간이 자연 지형에 개입해 형성한 도시 네트워크와 자연 공간 사이의 경계는 끊임없는 변화를 겪어왔다. 작은 산마다 경계의 유형은 물리적 존재와 그로 인해 생기는 변화 가능성에 따라 달라진다. 콘크리트 옹벽과 건물 같은 단단한 장벽에서 공원과 녹지 같은 유연한 전환 지대에 이르기까지 이런 경계는 사람들이 환경과 상호작용하고 주변 공간을 인식하는 방식에 영향을 준다.

와우산 (6)의 경우, 아파트 단지를 조성하기 위해 산비탈을 테라스처럼 평평하게 만드는 데 옹벽을 사용했는데, 이는 산의 단단하고 통과할 수 없는 가장자리를 강화하는 역할을 했다. 옹벽은 막다른 길을 돌아 공원을 통하는 길(와우시민아파트 붕괴 후 공원으로 변한 유명한 산비탈)로 방문객을 유도해 이 지역을 탐색하는 방식을 바꾸었다. **우장산** (20)에서는 보행자용 길이 여러 공원을 통과하기에 산 정상에 오르는 길이 다양하다. 하지만 경계가 더 모호해지는 경우도 있다. 예를 들어 **천장산** (1)에서는 산기슭 주변에 드문드문 지어진 대학교 캠퍼스와 다른 건물들로 인해 개방된 공간이 많아 보이지만, 정작 이런 공간에 접근할 수 있는 길은 거의 없다.

Vhile, when viewed from above, the unbuilt mountains ap-
ear as identifiable interruptions in the urban fabric, from the
round, the view of the mountains is in most cases obstructed.
idden behind tall buildings, their presence is often erased
om the mental map of the area. In some places, residents
ight live just a couple of blocks away from a hill or park with-
ut realising its existence due to visual and physical barriers.
or those in the know, these hidden hills can be secret gardens
r backyard mountains offering a unique and secluded escape
om the bustling city. Often remaining primarily pedestrian territories, he hills can be crossed via walking ails. Sometimes these paths are lanned, built, and maintained by the istrict authorities with clear indications on the map. A passerby might not dare to cross the rope handrail, which forms a soft boundary but a clear demarcation between the human world and that of the flora and fauna. In other cases, like in 3 봉화산 *Bonghwasan*, dozens of unpaved paths meander under the tree canopy, and the lack of clear limits invites exploration.

건물이 들어서지 않은 산은 위에서 내려다볼 때 도시의 구조물 사이에서 금방 눈에 띄지만, 땅에서 올려다볼 때는 산이 가려져 잘 보이지 않는다. 높은 건물 뒤에 가려진 산은 우리가 특정 지역을 떠올릴 때 산이 있다는 사실을 잊게 한다. 작은 산이나 공원에서 가까운 거리에 살면서도 시각적, 물리적으로 가려져 이곳에 산이나 공원이 있다는 사실을 잊기도 한다. 하지만 이를 잘 아는 사람들에게는 이렇게 숨겨진 작은 산이 번잡한 도시에서 특별하고 한적한 장소로 도피할 수 있는 비밀의 정원이나 뒷마당 역할을 한다. 대부분 보행 전용 구역으로 남아 있는 작은 산은 산책로를 이용해야 통과할 수 있다. 어떤 산책로는 지방자치단체가 계획, 조성하고 관리하기 때문에 지도에도 확실하게 표시된다. 이런 작은 산을 통과하는 사람은 인간과 동식물 구역 사이의 유연한 경계이자 확실한 구분선인 밧줄 난간을 넘을 수 없다. 봉화산 3 의 경우에는 지붕처럼 덮인 나무 아래로 비포장 길 수십 개가 구불구불 펼쳐지며 뚜렷한 경계가 없어서 이곳을 탐험하고 싶은 욕구까지 불러일으킨다.

3 봉화산 Bonghwasan
5 응봉산 Eungbongsan
12 국사봉 Guksabong
20 우장산 Ujangsan
KOREA NATIONAL UNIVERSITY OF ARTS 한국예술종합학교 석관동캠퍼스
UIREUNG (TOMB OF KING GYEONGJONG & QUEEN SEONUI) 의릉
KOREA INSTITUTE OF SCIENCE & TECHNOLOGY 한국과학기술연구원
GLOBAL KNOWLEDGE EXCHANGE AND DEVELOPMENT CENTER (GKEDC) 글로벌지식협력단지
KYUNGHEE UNIVERSITY 경희대학교서울캠퍼스
HONGNEUNG ARBOTEUM 홍릉시험림(홍릉숲)
KAIST SEOUL CAMPUS APARTMENT 한국과학기술원아파트
EAST SIDE OF CULTURAL CENTER 문화회관 동측주
KAIST COLLEGE OF BUSINESS KAIST 경영대학
KING SEJONG MEMORIAL HALL 세종대왕기념관
HONGNEUNG NEIGHBOURHOOD PARK 홍릉근린공원
YEONGHWIWON & SUNGINWON TOMBS 영휘원과 숭인원
1
The Nibbling of 1 천장산 Cheonjangsan Slope by the Institutions
천 장 산 1 의 경 사 를 서 서 히 잠 식 하 는 기 관 건 물 들
Car Access
Pedestrian Access
Dead End
Entrance
Building Boundaries
Artificial Green Space (Landscape)
Artificial Green Space (Urban)
Replanted Area

DOLMOE CHILDREN'S PARK
돌뫼어린이공원
UIREUNG ROYAL TOMB
서울 의릉
HANSOL APARTMENT
한솔아파트
IMUN CHILDREN LIBRARY
이문 어린이도서관
BETWEEN HONGNEUNG FOREST ARBOTEUM & KAIST SEOUL CAMPUS
홍릉숲과 KAIST 서울캠퍼스 사이
The Actual Resultant Voids in 1 천장산 Cheonjangsan
천장산 1 에 생겨난 실제 보이드
Mountain Area
Building Boundaries
Inaccesible Area
Car Access
Hiking Trails
Dead End
Entrance
3 봉화산 Bonghwasan
봉 화 산 3
12 국사봉 Guksabong
국 사 봉 12
6 와우산 Wausan
와 우 산 6
1
2
3
4
5

The presence of restricted military zones (군사시설물: military facilities [actual] and 특수지역: special area [land use]) adds another layer to the understanding of voids in the city. Mostly located on or in the vicinity of mountains, these areas are heavily fortified with fences, barbed wire, and surveillance cameras, creating inaccessible territories within the city. Ultimately, these are excluded areas for humans and buildings but not for birds, small animals, insects, and plants, and therefore they form precious areas for the ecosystem. The lack of accessibility is, in a way, the guarantee of the preservation of a certain biotope, turning these hidden spaces into resources for non-human life within the city. Some of these zones, however, have recently opened to the public, and the discovery of this new accessibility allows the expansion of a new buffer territory for the neighbourhood inhabitants.

On a daily basis, crossing the limits of the urban fabric allows the expansion or retraction of personal trajectories, opening areas of wilderness within the territory. This landscape of shifting boundaries and varying accessibility can suggest, to a certain extent, a gradient of freedom in the city.

출입이 제한된 군사 지역(실제 군사시설물)과 특수 지역(토지 용도)은 도시 내 보이드를 다른 방식으로 보게 한다. 주로 산 정상부나 그 주변에 위치한 이런 구역은 울타리, 철조망, 감시 카메라로 무장하고 있어 도시 안에서 접근 불가능한 구역을 형성한다. 궁극적으로 이 구역은 인간과 건물이 없는 장소이지만 새와 작은 동물, 곤충, 식물이 서식해 생태계가 보호되는 소중한 공간이다. 접근성이 떨어지는 이런 구역은 특정 생물군의 생태계를 보존하면서 도시 내 동식물을 위한 자원으로 전환된다. 하지만 그중에는 최근 대중에 개방된 곳도 있어서 이웃 주민을 위한 완충 지대가 확장되었다.

일상적으로 도시의 경계를 넘으면 개인의 경로가 확장되거나 축소되는데, 그 과정에서 야생 지역이 형성된다. 이런 경계 전환과 접근성 변화는 도시에서 자유의 증감으로 연결된다.

Overlapping Boundaries (Questioning Perception)

The overlapping of different boundary lines (topographic, solid/void, cadastral) questions our perception of these spaces. When does our connection to the mountain start? When the topography begins to change, and the slope becomes steeper? When we leave the urban fabric and see trees instead of buildings? When we cross the invisible cadastral line, the result of a succession of negotiations (moon village occupation (5) 응봉산 Eungbong-san, spiritual and institutional territories (Chapter 1) (1) 천장산 Cheonjangsan)? For some, the mountain stretches from their doorway with the donning of hiking shoes to wear up to the mountain's top. For others, the presence of inaccessible slopes at the back of their house allows them to connect and expand (physically or through the view) their inner territory to areas that they don't officially own. For some, the mountains are limiting obstacles, blocking the way between themselves and their destination, and therefore simply ignored.

경계의 중첩(인식 방식에 의문 제기하기)

다양한 경계선(지형선, 솔리드/보이드, 지적선)의 중첩은 이런 공간에 대한 우리의 인식 방식에 의문을 제기한다. 산과 연결되는 시점은 언제인가? 지형이 변하고 경사가 급해지는 때인가? 도시 구조를 벗어나면서 건물 대신 나무가 나타나는 때인가? 연이은 협상(달동네가 자리했던 응봉산 (5) 과 정신적, 제도적 지역(1장)인 천장산 (1))의 결과물인 보이지 않는 지적선을 넘어설 때인가? 누군가는 문을 나서서 등산화를 신는 순간부터 산과 연결된다. 다른 누군가에게는 집 뒤편에 오를 수 없는 산비탈이 있다는 건 자신의 내부 영역을 외부로 (물리적으로 또는 전망을 통해) 연결하고 확장할 수 있다는 의미다. 그리고 어떤 사람들에게는 자신과 목적지 사이의 길을 막는 제한된 장벽인 산은 그저 무시하는 대상이 되기도 한다.

These multiple lines sometimes overlap and sometimes diverge. But instead of searching for or defining a single outline, it is more intriguing to explore the interplay of these various lines. The perception of space is never fixed, and because it is never clearly defined, these buffers allow a level of freedom in the city. These shifting boundaries create an area in which apparently defined elements can move, allowing for multiple interpretations and possibilities. Going a step further, one can question the validity of looking for each mountain's individual outlines, and instead of seeing them as separated islands, we can view them as conglomerates of archipelagos, interconnected parts of a larger system.

이런 다양한 경계는 중첩되기도 하고 나누어지기도 한다. 하지만 확실한 경계선 하나를 찾는 대신, 이렇게 다양한 경계의 상호작용을 살피는 것이 더 흥미롭다. 공간을 인식하는 방법은 고정되어 있거나 명확하게 정의되지 않았기 때문에 이런 완충 지역은 도시 내에서 어느 정도 자유로움을 제공한다. 이렇게 변화하는 경계는 확정된 요소들이 약동하는 구역을 생성하며 다양한 해석과 가능성을 열어준다. 여기서 한 걸음 더 나아가, 모든 산의 고유한 윤곽선을 찾는 게 과연 필요한지 의문을 제기할 수 있으며, 그것들을 고립된 섬으로 보는 대신 더 큰 체계로 연결된 군도의 집합체로 볼 수 있다.

Chapter 3

3 장

Activities
경험

Human Use of the Voids
인간이 보이드를 활용하는 방식

일견 보이드처럼 보이는 이 작은 산들은 사실 빈 공간이 아니다. 여기에는 한때 달동네가 있었으며, 이곳들은 삼림 벌채 후 재조림 사업으로 복원돼 이제는 다양하게 활용되고 있다. 특히 도심지 건물 사이에 위치한 작은 산들은 광범위한 도시와 주변 지역에 혜택을 준다. 등산, 피크닉, 뒷마당 가꾸기부터 야외 운동, 맨발 걷기까지 다양한 활동이 가능한 확장된 여가 공간을 제공한다. 이런 가시적인 경관 아래에는 토지의 정확한 용도를 정의하는 보이지 않는 법적 경계가 존재한다. 이렇게 보이는 층과 보이지 않는 층이 결합하여 단순한 공원 이상의 기능을 한다. 주말 산책이나 잠깐의 휴식을 통해 주민들은 끊임없이 변하는 도시에서 그들의 역사와 정체성을 재발견한다.

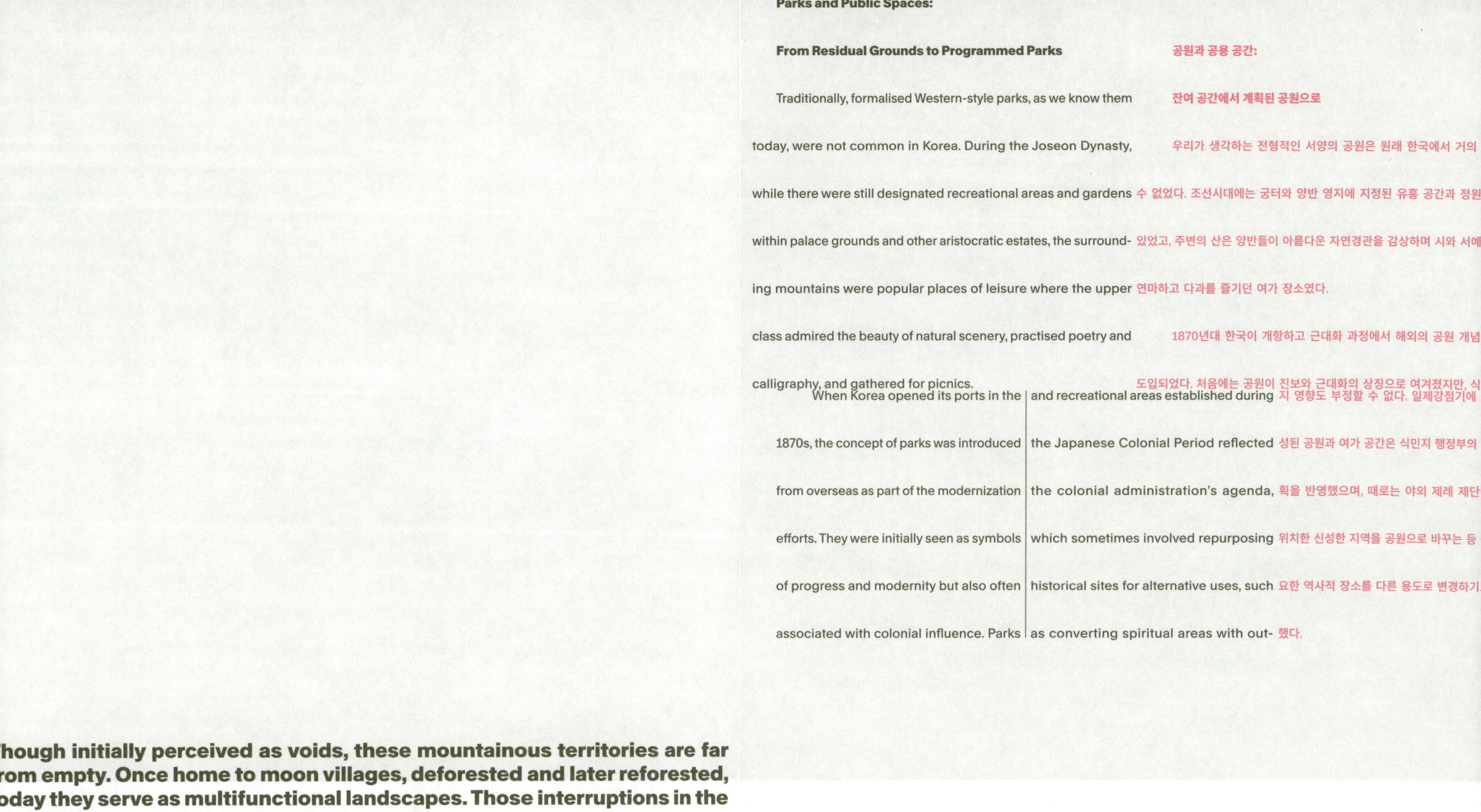

Parks and Public Spaces:

From Residual Grounds to Programmed Parks

Traditionally, formalised Western-style parks, as we know them today, were not common in Korea. During the Joseon Dynasty, while there were still designated recreational areas and gardens within palace grounds and other aristocratic estates, the surrounding mountains were popular places of leisure where the upper class admired the beauty of natural scenery, practised poetry and calligraphy, and gathered for picnics.

When Korea opened its ports in the 1870s, the concept of parks was introduced from overseas as part of the modernization efforts. They were initially seen as symbols of progress and modernity but also often associated with colonial influence. Parks and recreational areas established during the Japanese Colonial Period reflected the colonial administration's agenda, which sometimes involved repurposing historical sites for alternative uses, such as converting spiritual areas with outdoor ritual altars into public parks.

공원과 공용 공간:

잔여 공간에서 계획된 공원으로

우리가 생각하는 전형적인 서양의 공원은 원래 한국에서 거의 볼 수 없었다. 조선시대에는 궁터와 양반 영지에 지정된 유흥 공간과 정원이 있었고, 주변의 산은 양반들이 아름다운 자연경관을 감상하며 시와 서예를 연마하고 다과를 즐기던 여가 장소였다.

1870년대 한국이 개항하고 근대화 과정에서 해외의 공원 개념이 도입되었다. 처음에는 공원이 진보와 근대화의 상징으로 여겨졌지만, 식민지 영향도 부정할 수 없다. 일제강점기에 조성된 공원과 여가 공간은 식민지 행정부의 계획을 반영했으며, 때로는 야외 제례 제단이 위치한 신성한 지역을 공원으로 바꾸는 등 중요한 역사적 장소를 다른 용도로 변경하기도 했다.

Though initially perceived as voids, these mountainous territories are far from empty. Once home to moon villages, deforested and later reforested, today they serve as multifunctional landscapes. Those interruptions in the city's mass now accommodate functions that serve both the wider city and the surrounding neighbourhoods. Activities ranging from hiking, having picnics and planting backyard gardens to practicing outdoor fitness and barefoot walking also offer a territory for the expansion of leisure activities. Beneath this visible landscape lies a network of invisible legal boundaries that define the precise occupation of the land. These visible and invisible layers combine to make these mountains within the city far more than just parks; for a weekend walk or simply a moment's break, the mini-mountains allow the inhabitants to reconnect to their history and identity in an ever-changing city.

The introduction of Hanami, or cherry blossom contemplation, is an example of imported Japanese culture and influence during this time. According to records, city parks in Seoul during the colonial period were first created sporadically during the 1910s and 1920s, but increased to 7 parks in late 1928 and a total of 21 parks in the early 1930s, indicating a deliberate effort to develop urban green spaces. The reading of old maps reveals an early awareness and anticipation of the importance of defining not only the built but also the recreational zones for the city yet to be developed.

벚꽃을 감상하는 '하나미'는 일본에서 도입된 문화와 그 영향력을 보여주는 사례다. 기록에 따르면 식민지 기간 서울의 도시공원은 1910년대와 1920년대에 산발적으로 조성되기 시작했고, 1928년 말에는 7개로, 1930년대 초까지 총 21개로 증가했다. 도시의 녹지 공간을 개발하려는 의도적인 노력이 엿보인다. 옛날 지도를 보면, 도시 개발을 위해 건축물뿐 아니라 여가 공간이 중요하다는 사실에 대한 초기 인식과 기대감이 드러난다.

1 Park Planning Map of Seoul, 1920s
서울 공원 계획도, 1920년대

The Park Planning Map of Seoul

The Park Planning Map, simply dated 'After the liberation of Korea', was issued by the Land Survey Department of the Government General of the Republic of Korea. It gives both a vision of the ten existing parks and playgrounds at that time, but also enables us to see the strategic planning for the future city. Even at this early date, it already includes the different park categories such as Natural Parks (Protected Natural Spaces) allocated in the large mountainous areas (four to the north of the river and one to the south), as well as Nearby Park Areas, Planned Big Parks, and Sport Facilities homogeneously spread out over the unbuilt areas beyond the built city center (except one designated Nearby Park Area within the city center); and Park and Sport Facilities areas mainly located in the close vicinity of the old city center to the north of the river. Four mini-mountains: (2) 배봉산 *Baebongsan*, (7) 궁동 *Gungdong*, (11) 서달산 *Seodalsan*, and (12) 국사봉 *Guksabong* were part of the Protected Natural Space, and two mini-mountains (5) 응봉산 *Eungbongsan*, and part of (13) 용마산 *Yongmasan* were planned to be part of the Nearby Park Areas. As indicators, the seven light pink circles (each of them approximately the size of the old city) let us envision the expansion of the city in process. It is remarkable to see such an anticipation of the city's expansion and the early definition of void areas within what was, at the time, a vast and undefined territory. Careful examination shows that this copy was based on the 1920s map, itself successively edited during the Japanese Colonial Period. One might speculate that at the time of the Gyeongseongbu Park Planning Map's publication, the city itself was already much more developed than the built footprint that appears on the base map, but the precision of the base map and its detailed topography might have been a useful planning tool for the newly formed government.

서울 공원 계획도

제작 시기가 "대한민국 독립 이후"라고 표기된 「공원 계획도」는 조선총독부 토지조사과에서 발행했고, 당시 존재했던 10개 공원과 놀이터의 비전 그리고 미래 도시 계획을 보여준다. 이 지도는 이른 시점에도 이미 넓은 산악 지역(강북의 산 4개, 강남의 산 1개)에 지정된 자연공원(자연보호 지역)과 도심 너머 미개발 지역에 균일하게 분포된 인근 공원 구역(도심 내 지정된 인근 공원 구역 제외), 계획된 대형 공원, 체육 시설, 그리고 강북 구시가지 근처에 배정된 공원과 체육 시설 구역을 포함한다. 작은 산 4개(배봉산 (2), 궁동 (7), 서달산 (11), 국사봉 (12))는 자연보호 지역에 포함되고, 작은 산 2개(응봉산 (5), 용마산 (13))는 인근 공원 지역의 일부로 포함되었다. 연분홍색 원 7개(각 원은 대략 구시가지 크기)를 지표로 하면 도시 확장 과정을 상상해 볼 수 있다. 벌써 이른 시기임에도 용도 미정의 광활한 지역에 대해 도시 확장을 예상하고 보이드를 정의했다는 점이 놀랍다. 지도를 자세히 살펴보면, 일제강점기 동안 여러 차례 수정되었던 1920년대 지도의 사본임을 알 수 있다. 경성부 공원 계획도가 발행된 시기의 실제 모습은 기본 지도에 표시된 것보다 훨씬 더 발전했을 수도 있다. 하지만 기본 지도와 상세한 지형은 새로 구성된 정부의 계획에 유용한 자료로 쓰였을 것이 분명하다.

1
Seoul Metropolitan City Planning Park Change Plan, 1950
서울 도시계획 공원 변경 계획도(서울都市計畫公園變更計畫圖), 1950년ㄷ

The Seoul Metropolitan City Planning Park Change Plan

The Seoul Metropolitan Government's 1950s 'Seoul Metropolitan City Planning Park Change Plan,' shows the updated plan for parks and open green spaces. What strikes is the predominance of bright green surfaces and meanders that overlay the same 1:15,000 topographical base map left by the Japanese. Compared to the vague shapes of the earlier map, the precisely defined green areas follow the shape of contour lines, revealing an intention to integrate the natural topography into the planning of green spaces. This proposes a vision of the city aware of its natural assets—the four main Mountains, and the Hangang River, but also the smaller mountains such as (5) 응봉산 *Eungbongsan*, (6) 와우산 *Wausan*, (7) 궁동 *Gungdong*, (9) 노고산 *Nogosan*, (11) 서달산 *Seodalsan*, (12) 국사봉 *Guksabong*, and (13) 용마산 *Yongmasan*. The continuous network of green corridors that systematically connect the natural landmarks leads us to envision a city where the built environment finds its way between the natural spaces. The plan is developed on two sheets showing two successive phases. Compared to the first version, the second version shows an expansion of the mountainous areas of Bugaksan and the National Cemetery, while many of the small planned parks in the city have disappeared. Due to the chaos before the Korean War and its aftermath, Seoul's park vision shown in these maps could not be implemented. Although the creation of national parks such as the National Cemetery on (11) 서달산 *Seodalsan* was carried out, the Seoul districts densified quickly.

서울 도시계획 공원 변경 계획도

1950년대 서울시의 「서울 도시계획 공원 변경 계획도」는 공원 및 개방형 녹지를 업데이트한 계획으로, 일제강점기 때 사용한 1:15,000 지형도 위에 표시된 밝은 녹색 지역과 굽이진 모양의 곡류가 눈에 띈다. 이전 지도의 모호한 형태와 달리, 정확하게 표시된 녹지 공간은 지적선의 모양을 따르고 있어 자연 지형을 녹지 공간 계획에 통합하려는 의도를 볼 수 있다. 또한 주요 산 네 곳과 한강 그리고 응봉산(5), 와우산(6), 궁동(7), 노고산(9), 서달산(11), 국사봉(12), 용마산(13) 등 작은 산들을 자연 자산으로 인정하는 도시 비전을 확인할 수 있다. 자연 명소들을 체계적으로 연결하는 연속적인 녹색 그물망을 통해 건축물 구역이 자연환경 사이에서 자리잡는 도시를 상상하게 한다. 이 계획은 두 단계를 연속해서 진행하기 위해 두 가지 버전으로 발표되었다. 첫 번째 버전과 비교하면 두 번째 버전은 북악산과 국립묘지의 산악 지역이 확장 포함된 반면, 도시에 소규모로 계획한 공원이 많이 사라졌다. 한국전쟁 전후의 혼란스러운 상황 때문에 지도에 표시된 서울의 공원 계획은 실행되지 못했고, 서달산(11)의 국립묘지 등 국립공원이 조성되었지만 서울의 밀집도는 빠르게 증가했다.

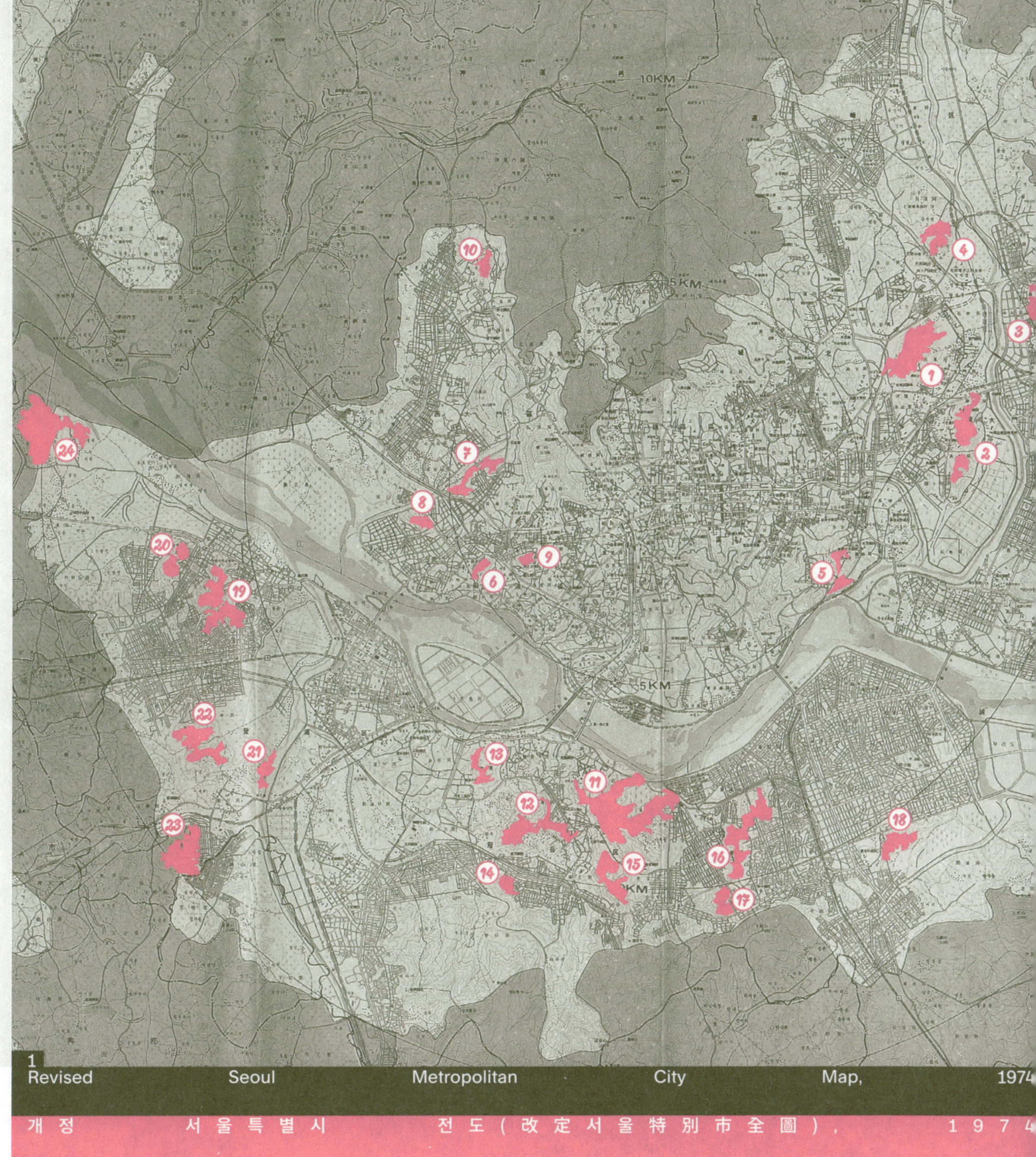

1 Revised Seoul Metropolitan City Map, 1974
개정 서울특별시 전도(改定서울特別市全圖), 1974

The Revised Seoul Metropolitan City Map

In 1973, the boundaries of Seoul expanded to almost ts current size of 605km. The 'Revised Seoul Metropolitan ;ity Map' (Revised Seoul Metropolitan Area) was published y Youngjin Culture Company in 1974. This map includes all he Seoul Metropolitan Districts as well as the road network highlighted in red) and subway lines, primarily indicating an rban planning purpose. Compared to previous maps, this ersion clearly shows the advancement of urban development nd land occupation on both sides of the river, including the Gangnam District. Besides serving urban development pur-oses, the map reflects a major step in the metropolitan vision: he delimitation of the "Greenbelt" (Development Restricted Zone) on the outskirts of Seoul.This is the result of an increasing ension between North and South Korea and also reflects the rowing concerns about environmental degradation and the oss of green spaces due to rapid industrialization and urban-zation in the region. Additionally, the eed to secure green buffer zones round military installations was a reoccupation during that time. Highlighted in green, these buffers einforce the clear contrast between he urbanized metropolitan area and he surrounding mountainous land. Pockets of barren land, in contrast to he grid network of transport, are still isible. Particularly noteworthy is the presence of all the selected mini-mountains surrounded by a relatively wider area of unbuilt land. However, compared to previous maps, the planning map doesn't include any schemes for public space, and without the few visible contour lines within the not yet developed areas, it's easy to perceive the city as built on flat ground. In fact, the grid development of Gangnam District can be misleading, as it is, in fact, situated on uneven topographic terrain.

개정 서울특별시 전도

1973년 서울시 경계는 거의 현재 수준인 605킬로미터로 확장되었고, 1974년 영진문화사에서 「개정 서울특별시 전도(개정 서울 도시 지역)」를 발행했다. 이 지도는 서울특별시 구역을 포함해 도로망(빨간색 표시)과 지하철 노선을 표시해 도시 계획의 목적을 나타낸다. 이전 지도들과 비교해 이 지도는 강남 지역을 포함한 한강 양쪽의 도시 개발과 토지 점유율의 변화를 확실히 보여준다. 도시 개발이라는 목적 외에도 이 지도는 수도권 비전의 중요 단계였던 서울 외곽의 '그린벨트(개발제한구역)'를 구분해 보인다. 그린벨트 구분은 남북 긴장이 고조된 결과물이자, 수도권 지역의 급격한 산업화와 도시화로 인해 환경 파괴와 녹지 손실에 대한 우려가 커졌음을 시사한다. 그리고 군사시설 주변에 완충 녹지대를 확보하는 것 역시 당시 주요 관심사였다. 녹색으로 강조된 이 완충 구역은 도시화된 수도권 지역과 주변 산지가 극명하게 대비되어 교통망과 대조적인 불모지가 군데군데 눈에 띈다. 특이한 점은 건물이 들어서지 않은 상당히 넓은 지역이 서울의 모든 작은 산을 둘러싸고 있다는 것이다. 하지만 이전 지도와 비교해 이 계획 지도에는 공용 공간에 대한 계획을 볼 수 없으며, 아직 미개발된 지역에 표시된 일부 지적선이 없으면 마치 도시가 평평한 지대에 건설된 것처럼 보이기 쉽다. 실제로 고르지 않은 지형에 위치한 강남의 격자형 개발은 오해를 불러일으킬 수 있다.

1
Distribution of Parks in Relation to Mini-Mountains in Seoul, 2010s
서울의 작은 산과 관련 공원의 분포도, 2010년대

Mini-Mountains 66	Urban Natural Park 68	Neighbourhood Park on Mountains 95	Neighbourhood Park on Flatlands 274

As urbanisation progressed and cities expanded, the role of nature and parks started to evolve in South Korean society. In particular, after the collapse of the Wau Citizen Apartment, new consideration for the natural landscape feature started to arise. A few apartment blocks originally built in the mountains were demolished and the locations returned to greenery (as previously mentioned in Chapter 2, this included the dynamiting of Namsan's surrounding apartments). A restriction on the height of buildings surrounding certain mountains was applied to preserve the scenic beauty of the area. Parks began to be seen not only as green spaces for environmental protection but also as essential components of urban living, serving the needs and interests of residents. To support this, the government started to launch a series of policies.

Between 1961 and 1979, parks were classified as children's parks, neighbourhood parks, urban natural parks, or cemetery parks. The classification of 'green spaces' into 'green buffer zones' and 'landscape zones' indicates a comprehensive approach to urban planning and environmental management. In addition, some heritage sites were turned into public parks. According to Article 38-2 of the National Land Planning and Utilization Act, the Korean government proposed constructing urban natural park areas to preserve or improve the natural environment and prevent environmental pollution and natural disasters in urban areas. It also aimed to improve the health, recreation, and aesthetic enjoyment of citizens. For example, the neighbourhood parks designed for the daily outdoor activities of the people living nearby. They are generally greater than one hectare in size, serving the people within a 500-meter radius. In Seoul, over 260 parks were implemented for such uses.

At the same time, due to the expansion of urban areas, the remaining unbuilt mini-mountains, which were once remote from living areas, became more accessible to residents. Several mini-mountains were relabelled as 'neighbourhood parks,' and their status shifted from residual empty ground to equipped spaces with facilities, providing residents with a place to spend leisure time in nature.

도시화에 따라 도시가 확장되면서 한국 사회에서 자연과 공원의 기능이 변해갔다. 특히 와우시민아파트 붕괴 이후 원래 자연환경을 중요하게 여기게 되었다. 산 위에 지어졌던 몇몇 아파트 단지가 철거된 후 산은 다시 녹지로 돌아갔으며 (2장에서 언급했듯이 남산 주변 외인 아파트 철거도 포함), 지역 자연환경을 보존하기 위해 특정 산 주변 건물 높이도 제한되었다. 공원은 환경보호를 위한 녹지 공간인 동시에 도시 주민의 필요와 관심을 충족하는 도시 생활의 필수 요소로 여겨졌고, 이를 지원하기 위해 정부는 새로운 정책을 마련했다.

1961–1979년에 공원은 어린이공원, 근린공원, 도시자연공원, 묘지공원으로 분류되었다. 녹지를 '완충 녹지 구역'와 '경관 녹지 구역'으로 분류한 것은 도시계획과 환경 관리를 통합적으로 관리하겠다는 접근법을 시사한다. 그리고 일부 문화유산 지역도 공원으로 변경되었다. 국토의 계획 및 이용에 관한 법률 제38조 2항에 따라, 대한민국 정부는 도시 지역의 자연환경을 보존하거나 개선하고 환경오염 및 자연재해를 예방하기 위해 도시에 자연공원 조성을 제안했다. 그리고 시민의 건강과 여가, 아름다운 경관을 향상하는 것도 목표로 삼았다. 예컨대 인근 주민의 일상적인 야외 활동을 촉진하고자 근린공원이 조성되었다. 근린공원은 일반적으로 1헥타르 이상의 면적으로 반경 500미터 이내 거주하는 주민들을 위한 공간이다. 서울에는 이런 용도로 사용되는 공원이 206개가 넘는다.

이와 동시에 도시가 확장되면서 생활권에서 멀리 위치해 개발되지 않았던 작은 산에도 주민들이 접근할 수 있게 되었다. 몇몇 작은 산은 근린공원으로 재분류되었고 잔여 공터에서 시설이 갖춰진 공간으로 변모하면서 주민들은 자연 속에서 여가를 즐길 수 있게 되었다.

Planting and forests

Nowadays, the remaining undeveloped summits that rise above the sea of construction allow inhabitants to breathe. This feeling of escape from the city is accentuated by the presence of trees. On these mountains, different species provide different spatial typologies and contribute to constructing a rich and varied experience. Offering varying degrees of shade during the summer and light filters during the winter, the vegetation in the parks evolves along with the seasonal rhythm, providing urban inhabitants with the opportunity to connect on a daily basis with the versatile capacity of nature. While this is also true for parks built on flat ground, the residual aspect of some mountains is specific to these buffers. As we saw earlier, the conversion of some mini-mountains into parks is relatively recent and based on connectivity with the neighbourhood, but some areas within the city are still kept relatively wild, mainly due to the lack of accessibility. This lack of accessibility is an advantage for the city as it allows the preservation of relatively untouched areas. However, upon closer inspection of the tree species, another layer of the history of these sites starts to unfold, and these spaces don't appear to be as 'natural' as they seem.

삼림 조성과 숲

오늘날 거대한 건물들 위로 솟아 있는 미개발 상태의 산 정상은 주민들에게 숨 쉴 수 있는 공간을 제공한다. 특히 나무가 무성할수록 도시에서 벗어나는 해방감이 증가하는데, 산에서는 다양한 수종이 공간 유형을 다채롭게 하고 풍부한 경험을 제공한다. 공원의 초목은 여름에는 다양한 음영의 그늘을 제공하고 겨울에는 빛을 통과시키는데, 계절에 따라 변화하면서 도시 주민에게 자연의 에너지를 매일 제공한다. 이런 특징은 평지에 조성된 공원에도 적용되지만, 산 속 공원으로 변모한 완충 지역에서 더욱 두드러진다. 앞서 보았듯, 작은 산이 공원으로 전환된 것은 최근 일이며 인근 지역과의 연결성에 기반한다. 하지만 일부 지역은 접근성이 떨어지기 때문에 자연 상태가 여전히 유지되고 있다. 덜 훼손된 야생 구역을 보존할 수 있어 오히려 도시에게는 장점이다. 하지만 수종을 자세히 살펴보면 숨겨진 또 다른 역사가 드러나는데, 이 지역들이 보이는 만큼 자연적이지 않다는 것을 알 수 있다.

During the Joseon Dynasty, the forest resources in Seoul, as in the rest of Korea, were abundant. As seen in Chapter 1, the areas dedicated to the logging of pine trees in the surrounding old city were made clear on a map. This allowed the maintenance of a balance between the forest and the built areas, favouring the overall ecosystem of the city.

However, at the end of the Joseon period, the deforestation of the territory started after the port opened and an agricultural promotion policy to develop the country's economy was implemented. Because of the lack of professional manpower, agricultural engineers, and capital, however, the agricultural expansion did not proceed smoothly.

Historical events such as the Japanese Colonial Period and the Korean War in the early 1900s were particularly damaging moments that led to widespread deforestation within and around the city. During the Japanese Colonial Period (1910–1945), natural resources were heavily exploited, resulting in a drastic reduction in forest stock from 700 million m³ to 200 million m³. The following Korean War (1950–1953) further caused severe forest cover loss as fuelwood was used as the primary source of energy for cooking and heating. Slash-and-burn agriculture increased rapidly, with the number of slash-and-burn farmers reaching almost two million, and accounting for 7% of the total population and 12% of farmers during that period.

In order to recover from these heavy loss of forest and the subsequent risk of soil erosion the Korean government launched various forest protection, reforestation, and erosion control projects through the establishment of the KFS in 1967; consequently, forest rehabilitation, achieved in over 20 years, was guided by the national forestry plan, which was revised every ten years.

조선시대에는 서울을 포함한 한국 전역에 산림 자원이 풍족했다. 1장에서 보았듯, 구시가지를 둘러싼 소나무 벌목 전용 지역이 지도에 정확히 표시되었다. 이런 정책을 통해 삼림과 개발 지역 사이의 균형이 유지될 수 있었고, 전반적인 도시 생태계에도 긍정적인 영향을 미쳤다.

그런데 조선 말기 개항 이후, 국가 경제를 발전시키기 위한 농업 진흥 정책이 시행되면서 산림 벌채가 시작되었다. 하지만 전문 인력과 농업 기술자, 자본의 부족으로 인해 농업 확장은 순조롭게 진행되지 못했다.

1900년대 초 일제강점기와 한국전쟁같이 역사적으로 중요한 시기는 특히 도시 내외부의 삼림이 광범위하게 벌채된 파괴적인 시간이었다. 일제강점기(1910–1945)에는 천연자원이 심각하게 착취되어, 삼림 자원이 7억 세제곱미터에서 2억 세제곱미터로 급격히 감소했다. 한국전쟁(1950–1953) 때는 땔나무가 요리와 난방의 주요 연료로 사용되어 심각한 삼림 손실을 초래했다. 화전 농업이 급속히 증가해 화전 농민 수가 거의 2백만 명에 달했는데, 당시 전체 인구의 7퍼센트, 농민의 12퍼센트를 차지했다.

이렇게 심각한 삼림 손실과 그로 인한 토양 침식의 위험을 회복하기 위해, 정부는 1967년 한국산림청(KFS)을 설립해 삼림 보호, 재조림 사업, 침식 방지 사업을 다양하게 펼쳤다. 그 결과, 10년마다 국가 산림 계획을 개정하면서 20년 만에 산림 복원을 달성했다.

1 Namsan Erosion Control Construction

남산의 침식 방지 공사

To address the large scale deforestation, the Government launched six successive Forest Rehabilitation Plans. The first phases (1973–1978) had as their primary goal to rehabilitate Korea's denuded landscape with fast-growing trees, focussing on the promotion of a national reforestation movement in which all citizens could participate, the expeditation of reforestation by planting fast-growing trees (larch, poplars, black locusts, etc.), the creation of fuel forests, and the rebuilding slash-and-burn fields. The Second plan (1979–1987) emphasized the economic role of forests, continuing reforestation efforts initiated from the previous 10 year plan and also aiming to bring soil erosion under control. The third plan, named the Forest Resource Plan (1988–1997), aimed to develop income streams and improve the public service function of forests to maximize forest utilization. The Fourth Forest Basic Plan (1998–2007), which prioritized sustainable forest management (SFM) was first introduced as the primary direction for forest management. The major objective of this plan included the development of forest resource management systems to maintain sustainable forest ecosystems and contribute to the economy. Beginning in 2008, the fifth National Forest Basic Plan (2008–2017) was established to develop forest education programs, recreational services, and healing opportunities as civic demand for forest use grew. The current sixth National Forest Basic Plan is the nation's first twenty-year plan (2017–2037), designed to foster healthy and valuable forests that promote the public interest, provide quality forestry jobs, and help North Korea restore its devastated forests in preparation for the future reunification of the Korean peninsula.

대규모 산림 벌채를 해결하기 위해 정부는 여섯 단계의 산림 녹화 사업 계획을 발표했다. 제1차 계획(1973–1978)은 성장이 빠른 나무를 심어 황폐화된 산의 복원을 목표로 삼았고, 전 국민이 참여하는 국가 차원의 나무 심기 운동에 집중했다. 특히 속성수(낙엽송, 포플러, 아카시아나무) 조림으로 사업을 가속화하며, 연료용 삼림을 조성하고 화전 농지를 복원했다. 제2차 계획(1979–1987)은 산림의 경제적인 역할을 강조해 이전 10개년 계획으로 시작된 재조림 사업을 지속하면서 토양침식 관리에 중점을 두었다. 제3차 삼림자원계획(1988–1997)은 수익원을 개발하고 산림의 공공서비스 기능을 개선해 산림 활용을 극대화하고자 했다. 제4차 산림기본계획(1998–2007)은 산림 관리의 주요 방향을 지속 가능한 산림 관리(SFM)로 처음 설정했다. 주요 목표는 지속 가능한 산림 생태계를 유지하고 경제에 기여하도록 산림자원을 관리하는 시스템을 구축하는 것이었다. 2008년부터는 산림을 이용하는 시민의 수요가 증가하면서, 제5차 산림기본계획(2008–2017)을 통해 산림 교육 프로그램, 휴양 서비스, 힐링 체험을 개발했다. 현재 진행 중인 제6차 산림기본계획(2017–2037)은 최초의 20년 계획으로, 공익을 증진하고 임업 관련 양질의 일자리를 제공하며, 미래 한반도 통일에 대비해 황폐해진 북한 산림의 복구를 지원하고 건강하고 가치 있는 숲을 만드는 것이 목표이다.

Today, South Korea boasts 6.3 million hectares of mountainous forest areas, encompassing 63% of the total national area. According to the Ministry of Environment, around 30% (31.9%) of the Seoul Metropolitan Area is occupied by forest and open space, and 8.1% is river, stream, or wetland. The rest consists of residential areas of around 18.9%, 13% mixed residential and business area, 10.5% transportation facilities, 5.9% commercial and business areas, and 5.1% public facility areas. In addition, the specific type of land use labelled 'inaccessible area' (Chapter 2), occupied 14,490,734 m², 2.4% of Seoul's city area (608,318,759 m²). According to the *KFSS Korean Forestry and forestry statistical yearbook*, there was a slight decrease between 2017 and 2022 (15,387 ha in 2017 and 15,362 in 2022) however the volume of tree cover is increasing.

현재 한국은 전체 국토 면적의 63퍼센트를 차지하는 630만 헥타르의
림 지역으로 덮여 있다. 환경부에 따르면 수도권의 약 30퍼센트(31.9퍼
트)가 숲과 개방 공간이며, 8.1퍼센트가 강과 하천, 습지다. 나머지 면적
주거 지역 18.9퍼센트, 혼합 주거 및 상업 지역 13퍼센트, 교통 시설 10.
퍼센트, 상업 및 사업 지역 5.9퍼센트, 공공 시설 지역 5.1퍼센트를 포함
다. 또한 '접근 금지 지역'(2장 참조)으로 표시된 특정 토지이용 지역은
울시 면적(608,318,759제곱미터)의 2.4퍼센트인 14,490,734제곱미터를 차지한다. 『한국 산림 및 산림 통계 연감(KFSS)』에 따르면 2017년과 2022년 사이에 소폭 감소했지만(2017년 15,387헥타르, 2022년 15,362헥타르) 숲으로 덮인 면적은 전체적으로 증가하고 있다.

Observing the distribution of tree species across each mini-mountain, one can almost trace the distinct history of each site. The mixed and varied species pattern on 24 개화산 Gaekwasan suggests relatively untouched ground, perhaps due to its location at the edge of the city; this contrasts with the monoculture coverage of 6 와우산 Wausan, where (Chapter 2) apartment blocks used to occupy the slopes. Recalling the dense coverage of the hills by the moon villages on 5 응봉산 Eungbongsan and 12 국사봉 Guksabong suggests that the trees we see today are relatively young. We also notice efforts to increase the variety of tree species on 20 우장산 Ujangsan, which aligns with the richness of activities available there, in contrast to the more uniform planting on 7 궁동 Gungdong, which resembles an urban buffer rather than a designed park.

작은 산에 분포된 수종을 관찰하면 각 지역의 독특한 역사를 추적할 수 있다. 개화산 24 의 다양하고 혼합된 수종은 비교적 훼손되지 않았음을 보여주는데, 아마도 도시 외곽에 위치했기 때문인 듯하다. 이런 특징은 (2장에서 보았듯이) 아파트 단지가 경사면을 뒤덮었던 와우산 6 의 단일 수종과 대조된다. 응봉산 5 과 국사봉 12 을 빼곡하게 채웠던 달동네를 생각하면, 오늘날 우리가 보는 나무는 상대적으로 어린 나무라는 걸 짐작할 수 있다. 우장산 20 에서는 수종의 다양성을 확보하려는 노력이 돋보이며, 그곳에서 제공되는 다양한 활동과 잘 어울러진다. 반면, 궁동 7 은 균일한 수종이 눈에 띄는데, 계획 공원이 아니라 도시 완충 지대와 유사한 모습을 보인다.

5 응봉산 Eungbongsan

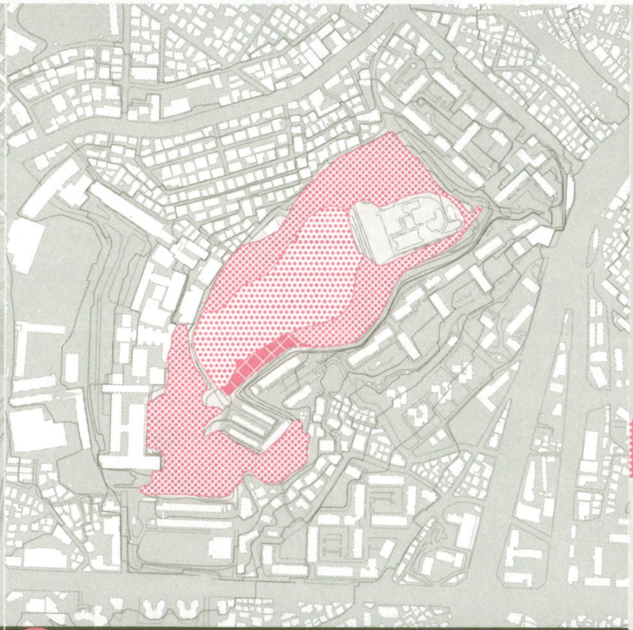

6 와우산 Wausan

20 우장산 Ujangsan

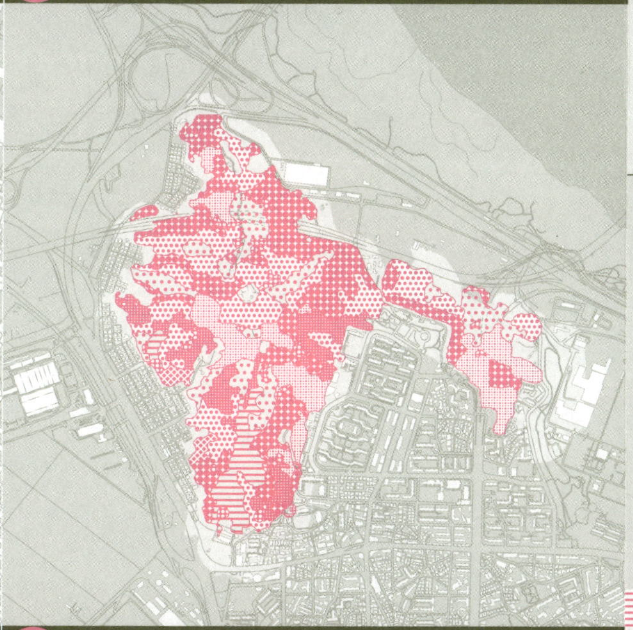

24 개화산 Gaehwasan

편백나무 Cypress	상수리나무 Sawtooth Oak	아카시아나무 Black Locust
리기다소나무 Rigida Pine	기타 참나무류 Other Oaks	정자나무 Shade Trees
소나무 Pine	제지 Paper Making	관목덤불 Shrub Bushes
잣나무 Korean Pine	소사나무 Silky Hornbeam	삼나무 Cedar
느티나무 Zelkova	기타 침엽수 Other Conifers	기타 활엽수 Other Hardwood/ Broadleaf

아카시아나무 Black Locust

Native or non native (to South Korea): Non-native (native to North America)
Height and dimensions: 25 m
Growth speed: 1–2 m per year

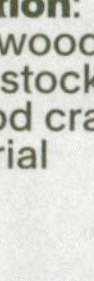

Function:
- Firewood source
- Livestock feeding
- Wood crafting material

삼나무 Cedar

Native or non native (to South Korea): Non-native (native to Japan and China)
Height and dimensions: 40 m, 1–2 in diameter
Growth speed: 2–3 m per year

Function:
- Building material
- Wooden crafts (frame making)

벚나무 Cherry Blossom

Native or non native (to South Korea): Native
Height and dimensions: 10–20 m
Growth speed: 2.5 m after 4–5 years

Function:
- Herb for food
- Wood craftin material

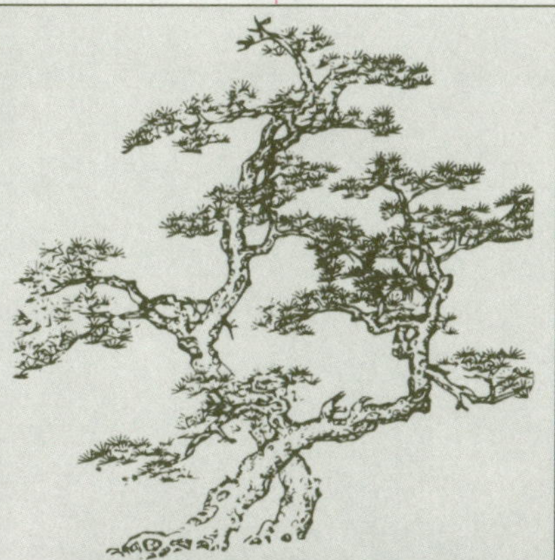

소나무Pine

Native or non native (to South Korea): Native
Height and dimensions: 35 m, 1.8 in diameter
Growth Speed: 30 cm yearly

Function:
- Building material
- Ink
- Pesticide
- Rice cake making
- Tea

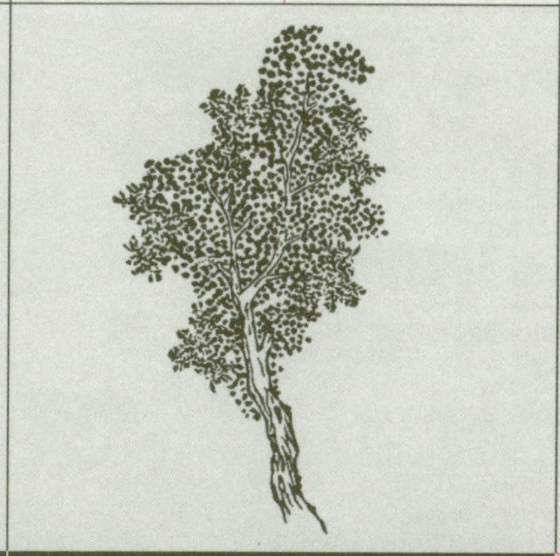

포플러 Poplar Tree

Native or non native (to South Korea): Non-native (native to Europe)
Height and dimensions: 30 m
Growth Speed: 30 cm yearly on average
Note: It has strong cold resistance, so it grows well anywhere in the country, and can especially be found in river basins, rice fields, and field banks. It has a high high requirement for sunlight and has strong resistance to moisture, sea breezes, and air pollution

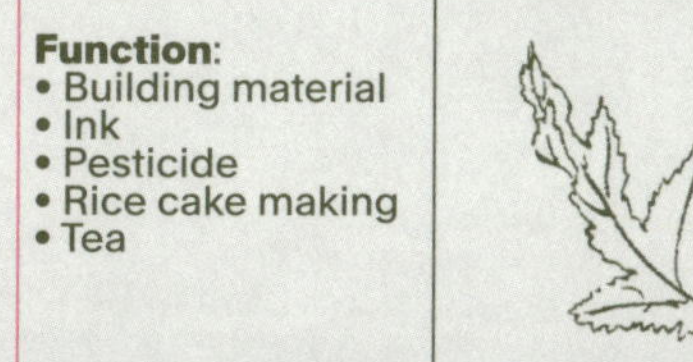

Function:
- Building material
- Cooking aid (BBQ)
- Food packaging
- Livestock food (poplar leaf)

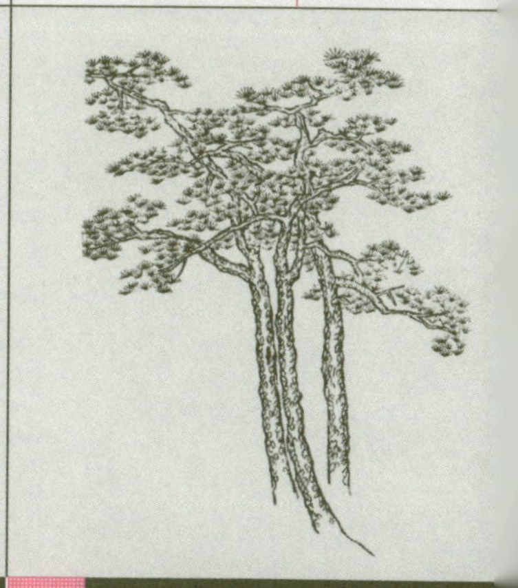

리기다소나무 Rigida Pine

Native or non native (to South Korea): Non-native (native to North America)
Height and dimensions: 25 m, 1 m in diameter
Growth Speed: 30 cm yearly on average

Function:
- Afforestation
- Building mat
- Erosion cont

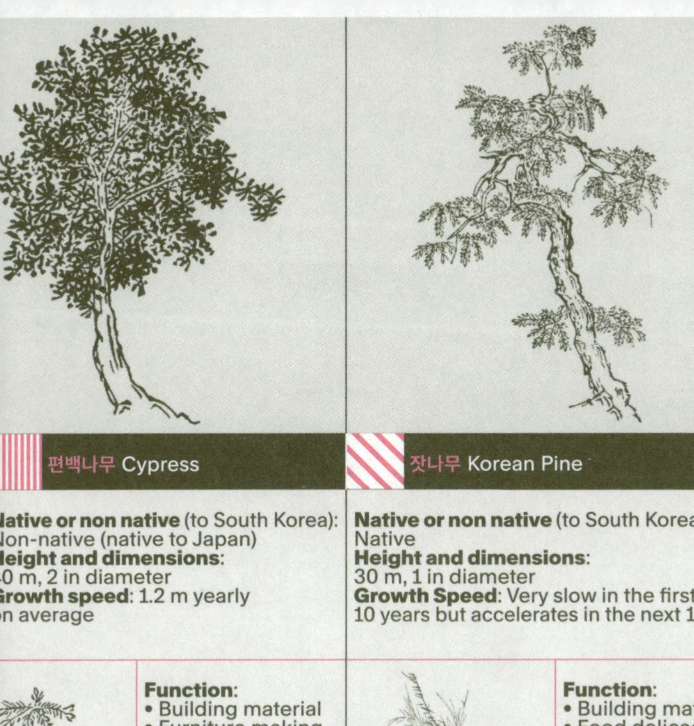

편백나무 Cypress

Native or non native (to South Korea):
Non-native (native to Japan)
Height and dimensions:
.0 m, 2 in diameter
Growth speed: 1.2 m yearly
on average

잣나무 Korean Pine

Native or non native (to South Korea):
Native
Height and dimensions:
30 m, 1 in diameter
Growth Speed: Very slow in the first
10 years but accelerates in the next 10 years

Function:
- Building material
- Furniture making

Function:
- Building material
- Food delicacy (porridge)
- Wooden crafts

상수리나무 Sawtooth Oak

Native or non native (to South Korea):
Native
Height and dimensions:
.0–25 m, 1 m in diameter
Growth Speed: 90 cm yearly
on average

느티나무 Zelkova

Native or non native (to South Korea):
Native
Height and dimensions:
26 m
Growth Speed: 30–60 cm yearly
on average

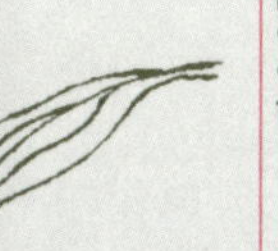

Function:
- Charcoal making
- Herbicides for farming

Function:
- Agricultural tools
- Building material
- Furniture making

Functions

Sanseujang 산스장

(Mountain's Facilities)

Within these enclaves, a diverse range of activities animates the mini-mountains and connects them to the surrounding neighbourhoods. They are called Sanseujang 산스장 (mountain facilities) as opposed to Gongseujang 공스장 (park facilities). These activities are supported by designated areas such as playgrounds, outdoor fitness zones, and viewing platforms—features commonly found on many mountains. The layout of these spaces is carefully designed, with open-air grounds that are often shaded by tree canopies and bordered by natural elements like shrubs and bushes. Unlike indoor sports centres, these outdoor spaces serve not only as places for physical exercise but also as social hubs where residents can connect with one another and experience the changing seasons.

기능

산스장(운동기구 시설)

작은 산에서 할 수 있는 다양한 활동은 산에 활기를 불어넣고 주변 지역과 연결 고리를 형성한다. 산에 있는 이런 시설은 '공스장(공원의 운동기구 시설)'과 비교해 '산스장(산의 운동기구 시설)'이라고 불린다. 대부분의 산에서 흔하게 볼 수 있는 놀이터, 야외 운동 구역, 전망대 등 지정된 공간에서 다양한 활동을 할 수 있으며, 나무 그늘과 덤불 같은 자연환경으로 둘러싸이도록 세심하게 설계된 점이 특징이다. 실내 스포츠센터와 달리, 이런 야외 공간은 주민들이 체육 활동도 하고 서로 소통하며 계절의 변화를 느낄 수 있는 사회적 허브 기능도 담당한다.

Most of the activities are organised and managed by the local districts, but upon closer observation, there appear to be two types of outdoor setups: the standard ones provided by the government, and those created through individual initiatives. In (12) 국사봉 Guksabong, for example, one of the fitness areas at the mountain's summit stands out because of its unstandardised equipment, mirrors, and clocks—likely arranged by the users themselves. This informal setup reflects the close connection between the users and their neighbourhood.

대부분의 활동은 지방자치단체에서 조직하고 관리하지만, 자세히 살펴보면 야외 활동에는 두 가지 유형이 있는 듯하다. 하나는 정부에서 제공하는 표준 활동이고, 다른 하나는 개인이 주도하는 활동이다. 예를 들어 국사봉(12) 산 정상의 운동 장소 중 하나가 눈에 띄는데, 특별한 장비와 거울, 시계 등이 놓인 것으로 보아 사용자들이 직접 설치한 것처럼 보인다. 이런 자유로운 공간에는 사용자와 주변 환경의 밀접한 관계가 반영된다.

In other places, activities are deeply tied to the historical context of the area. The rose garden on 5 응봉산 *Eungbongsan*, for instance, was cultivated on land that was once part of a moon village, and it exists today as the result of collaboration between the local government and residents. Similarly, in 18 매봉산 *Maebongsan*, a small, semi-hidden badminton court, apparently initially set up by a group of older women, can be found nestled between the bushes. Situated in the heart of the highly developed Gangnam District, this space surprises with its sensitive, user-driven design, reflecting a deep understanding of the place by those who frequent it, rather than top-down district planning. A similar example can be seen on 3 봉화산 *Bonghwasan*, where a table tennis area and its surrounding nets were clearly arranged by the users themselves. As stipulated in the Urban Parks and Green Areas Act, the law allows anyone to propose an activity, creating a healthy balance between resident initiatives and government contributions.

산에서 일어나는 활동 중 지역의 역사적 배경과 깊이 연관된 경우도 있다. 예를 들어 응봉산 5 의 장미 정원은 이전 달동네에서 재배되기 시작해, 지방자치단체와 주민들의 협력으로 지금까지 유지되고 있다. 유사한 예로 매봉산 18 에는 작은 '할머니 배드민턴장'이 덤불 속에 숨겨져 있는데, 할머니들이 설치한 것으로 알려졌다. 강남 중심부에 자리 잡은 배드민턴장은 지방자치단체의 하향식 계획이 아닌 사용자 중심의 세심한 설계로, 이곳을 자주 사용하는 주민들이 장소를 얼마나 잘 이해하고 있는지 보여준다. 봉화산 3 에서도 사용자들이 탁구장과 주변 네트를 직접 배치한 것이 분명한 유사한 공간을 볼 수 있다. '도시공원 및 녹지 등에 관한 법률'에 명시된 것처럼 누구나 활동을 제안할 수 있기에, 주민이 주도하는 활동과 정부의 지원금 사이에 건전한 균형이 유지된다.

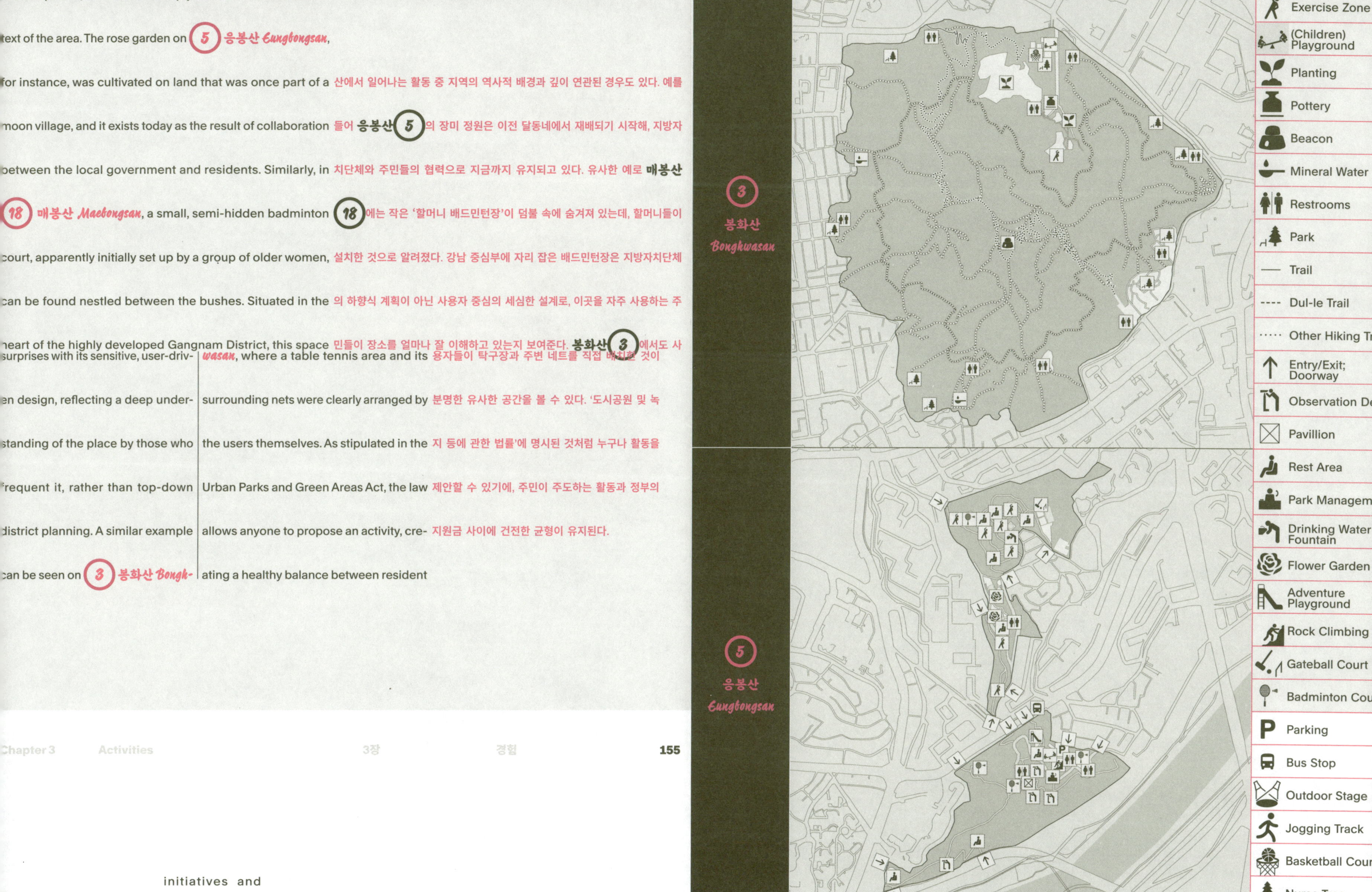

Management Office
Forest Fire Extinguisher
Playground
Metro Station
6
7
12
국사봉
Guksabong
16
서리풀
Seoripul
Exercise Zone
(Children) Playground
Planting
Pottery
Beacon
Mineral Water
Restrooms
Park
Trail
Dul-le Trail
Other Hiking Tra
Entry/Exit; Doorway
Observation De
Pavillion
Rest Area
Park Manageme
Drinking Water Fountain
Flower Garden
Adventure Playground
Rock Climbing
Gateball Court
Badminton Cou
Parking
Bus Stop
Outdoor Stage
Jogging Track
Basketball Cour
Nurse Tree

Management Office
Forest Fire Extinguisher
Playground
Metro Station
18
배봉산
Baebongsan
20
우장산
Ujangsan
GEOMDEOKSAN PEAK
검덕산
AMPHITHEATRE
공연장
COMMUNITY CENTRE
강서구민회관
ARCHERY RANGE
CHILDREN'S FOREST PARK
RED CLAY PATH HEALING CENTRE
UJANGSAN PEAK
우장산
SPORT COURTS
운동장
KOREA POLYTECHNIC UNIVERSITY
한국폴리텍대학
우 장 산

he intensity of use varies from one mini-mountain to another, especially in areas surrounded by large apartment complexes. n 20 우장산 *Ujangsan*, for instance, these pockets of activities include: an archery training stall, football, tennis, basketball, badminton court, fitness installations, resting pavilion, pring water collection points, playgrounds, forest adventure hildren's playground, an outdoor stage, community center, oetry trails, a forest library, and a healing center. The density f activity almost mirrors the built density of the surroundings. nd on summer evenings, it feels as though the whole community—from infants to the elderly—gathers on the walking rails that circle the mountain, providing a much-needed reathing zone for the entire populations of the residential owers. This over-programming attitude can make us question he real ecological values of these "green" spaces. But thanks o the lack of accessibility, some of the mini-mountains are till preserved as natural biotopes; a place for non-humans and a shelter for plants and animals. Here and there, one can also observe that even the humans do not take the official path designed by the government—as if travelling across a wild, natural territory—indicating a higher level of agency in their interpretation of the landscape.

특히 대규모 아파트 단지로 둘러싸인 지역에서는 작은 산을 활용할 수 있는 방법이 많다. 예를 들어 **우장산** 20 에서는 양궁, 축구, 테니스, 농구, 배드민턴 코트, 피트니스 시설, 휴식 정자, 약수터, 놀이터, 어린이 숲 체험장, 야외 무대, 커뮤니티 센터, 시 산책로, 숲 도서관, 힐링 센터 등 다양한 활동이 가능하다. 이렇게 풍부한 활동은 주변 건축물의 밀집도를 반영한다. 특히 여름 저녁에는 유아부터 노인까지 지역 주민 전체가 산을 둘러싼 산책로에 모이는 듯한 풍경이 연출되고, 주거 단지에 사는 주민들에게 절실할 휴식 공간을 제공한다. 이렇게 집중된 활동 프로그램은 '녹지' 자체의 생태적 가치에 의문을 제기하게 한다. 하지만 작은 산의 일부 구역은 접근성이 떨어지기 때문에 여전히 인간을 배제한 동식물의 쉼터, 천연 생태 서식지로 보존된다. 관련 지역을 둘러보면 사람들이 정부에서 조성한 공식 산책로를 따르지 않고 야생 자연 구역을 탐험하듯 가로지르는 모습이 관찰되는데, 경관을 해석하는 데 높은 자율성을 중시함을 알 수 있다.

Urban Mountain or Public Space?

도시의 산인가, 공용 공간인가

The need for natural public spaces like these is undeniable, but what makes these places unique compared to typical parks? While they may now seem well-coordinated, offering a similar range of activities managed by the district, each one has its own history. The ease with which these natural spaces now serve their neighbourhoods belies the fact that they were once under threat. Many of these areas were once slums or could have been developed into apartment blocks, so the remnants of their history make them distinctive.

자연적인 공용 공간의 필요성은 부인할 수 없지만, 전형적인 공원과 비교해 특별한 이유는 무엇일까? 지금은 지방자치단체가 자연 공간을 관리하며 비슷한 활동을 제공하는 것처럼 보이지만 역사는 각기 다르다. 현재 자연 공간은 지역 주민 곁에서 잘 기능하고 있지만, 사실 존재 자체가 위협받는 시기도 있었다. 많은 지역이 한때 빈민가였거나 아파트 단지로 개발될 가능성이 있었기에 역사적으로 특별한 의미가 있다.

Nestled on the slopes behind residential blocks, small portions of mountains resemble backyard gardens, where flowers and vegetables are cultivated by the nearby inhabitants. Occasionally, a hut or village type of house serves as a reminder that not long ago, these hills were surrounded by fields. At large-scale, one notable example is a pear orchard located on the northern side of 3 봉화산 *Bonghwasan*. Covering an area of around 20,000 m², hundreds of pear trees homogeneously spread out to offer shade to hundreds of tiny garden plots.

주거 단지 뒤편 경사면을 차지하는 산 일부 구역은 인근 주민들이 꽃과 채소를 재배하는 뒷마당처럼 보인다. 가끔 볼 수 있는 오두막이나 시골집 같은 건물은 얼마 전만 해도 이 언덕들이 들판으로 둘러싸여 있었음을 상기시킨다. 주목할 만한 예시는 **봉화산** 3 북쪽에 위치한 배 과수원으로, 수백 그루의 배나무가 약 20,000제곱미터 면적을 균일하게 차지하고 작은 정원 수백 개에 그늘을 제공한다.

Developed on a bowl-shaped slope and irrigated by a small stream, this communal garden is the result of years of patient cultivation. Although the orchard is managed by a single farmer, each of the parcels below the trees is maintained by a different gardener. Scattered across the hillside, small shelters built by individual gardeners showcase their creativity, each imbued with a personal touch. Along the pathways, an ingenious water-bucket system is available for all to share, ensuring that the gardens are well-watered. At the base of the hill, larger pavilions made of combined umbrellas provide space for the community to gather.

수년 동안 인내하며 재배한 이 공동 정원은 움푹 팬 모양의 경사면에 있고 작은 개울로 관개 시스템을 갖추었다. 과수원은 농부 한 명이 관리하지만, 각 나무 아래 개별 공간은 다른 정원사들이 관리한다. 언덕을 따라 드문드문 지은 작은 쉼터는 개별 정원사의 개성이 담긴 창의성을 보여준다. 길을 따라 설치된 기발한 물통 시스템은 모든 사람이 충분한 양의 물을 공급받을 수 있게 한다. 언덕 아래에는 우산 여러 개로 만든 큰 정자가 있는데, 지역 주민이 모일 수 있는 공간을 제공한다.

12 국사봉 Guksabong

This place is captivating because it embodies the perfect example of a public space that is shared, maintained, and cherished by the neighbourhood's residents. However, at the time of observation, the future of this garden hangs in the balance. Billboards indicate that the garden may soon be erased and replaced by a more formal public space. The ambiguity of access—part of its allure—may deter some from entering, but wouldn't the charm and unique character of this hidden garden be lost forever if replaced by a more clearly defined, regulated public space?

Fortunately, unlike parks on flat terrain, these mountainous areas hide pockets of inaccessible land that maintain a sense of wilderness. Despite the fact that many of these urban forests are relatively young, they hold an air of mystery. These untouched pockets of land allow for the coexistence of wildlife within the city. Across these green spaces, we can observe a gradient of artificiality. At the base of the mountains, highly manicured gardens are common, but when climbing higher, the landscape more closely resembles a forest. The topography and relative inaccessibility of these areas create a unique blend of city and wilderness within these micro territories.

이런 장소는 지역 주민이 공유하며 소중하게 여기는 공용 공간의 완벽한 사례이기에 더욱 매력적이다. 하지만 현재 이 정원의 미래는 불투명하다. 정원을 공식적인 공용 공간으로 대체할 수 있다는 공지가 발표되었기 때문이다. 물론 접근성이 모호해 어떤 사람은 출입하기 주저할 수도 있지만, 앞으로 더 확실하게 정의된 공용 공간으로 대체된다면 이 숨겨진 정원의 매력과 독창성이 계속 유지될지는 의문이다.

평지에 있는 공원과 달리, 다행히 산악 지역은 접근하기 어려운 구역이 몇 군데 있어서 야생 상태를 유지한다. 도시의 숲은 대부분 상대적으로 젊지만 신비로운 분위기를 띤다. 사람의 손길이 닿지 않은 구역은 도시에서 야생 동식물과의 공존을 가능하게 한다. 이런 녹지 공간에서도 인공적인 개발의 다양성을 관찰할 수 있는데, 산기슭에는 잘 관리된 정원이 많지만 위로 오를수록 경관은 더욱 숲에 가깝게 변한다. 이런 구역은 지형과 접근성 제한 때문에 도시와 야생 자연이 독특하게 혼합된다.

12

국사봉

Guksabong

1

국 사 봉 12

Hwangtogil 황토길

Among the variety of activities found in these mountain enclaves, one is particularly surprising—walking barefoot on the earth. Along red dirt paths, amidst tangled roots and rocks, some wander barefoot, softly tapping the mountain's surface with the soles of their feet. It's not unusual to see a pair of shoes left behind, waiting for their owner, who has set off to explore the delicate warmth of the mountain's skin through direct contact. This practice of walking barefoot, which has gained popularity recently, seems to offer more than just a physical experience; it has become a communal activity, as seen in the growing number of "healing" centres and dedicated "Red Clay Trail" or Hwangtogil 황토길 either already present or under construction in the mountains. These Hwangtogil refer to Hwangto (황토), the Korean term for loess soil, a specific red coloured soil, which contains high levels of potassium chloride and calcium. It is sometimes called a 'living soil' for its medical effects.

의식과 관행

황토길

작은 산에서 볼 수 있는 다양한 활동 중 특히 놀라운 것은 바로 맨발로 걷는 체험이다. 사람들은 엉킨 나무뿌리와 돌이 있는 황토길을 맨발로 부드럽게 밟으며 걷는다. 산 지면의 온기를 체험하러 떠난 주인을 기다리는 신발들을 드물지 않게 볼 수 있다. 최근 성행하는 맨발 걷기는 단순한 체험 이상의 공동 활동이 되었고, 산에는 조성 중이거나 이미 조성된 힐링 센터와 전용 황토길 수가 늘어나고 있다. 황토길의 황토는 염화칼륨과 칼슘이 다량 함유된 독특한 붉은색 토양을 지칭하며, 의학적 효과로 인해 '살아 있는 흙'이라고 알려졌다.

In (20) 우장산 *Ujangsan*, for instance, the Hwangtogil 황토길 is lined with a low stone wall, offering a place for walkers to pause and rest. Before entering, removing one's shoes is, of course, mandatory, with shelves at the entrance resembling those seen outside public baths. After a few rounds of walking along the mud path, and perhaps soaking their feet in a puddle, participants can wash off in nearby fountains, which evoke the charm of old-fashioned laundry stations. Plastic buckets are scattered around for convenience, turning this simple act of washing into yet another moment to interact with others and share in the healing power of the earth beneath their feet.

What may appear to be a simple mountain ritual carries a deeper significance—a profound reconnection to nature and a reinforcement of local identity. In a cityscape dominated by concrete and tarmac, where the ground is often concealed beneath layers of modernity, these mini-mountains provide a rare and ancestral link to the natural world. Walking barefoot, free from the constraints of urban life, offers not just a return to the earth, but a return to something older, something shared. Through this practice, tradition and identity are not relics of the past but are instead lived experiences, woven together through a collective sense of freedom and a renewed bond with the land.

예를 들어 **우장산 (20)** 의 황토길은 낮은 돌담을 따라 이어져 산책하는 사람이 잠시 휴식할 수 있는 공간을 제공한다. 물론 입장하기 전에 신발은 당연히 벗어야 해서 황토길 입구에는 대중목욕탕 신발장과 비슷한 선반이 준비되어 있다. 황토길을 따라 몇 바퀴 돌면서 발이 젖으면 근처 수돗가에서 발을 씻으면서 옛날 빨래터 기분을 낼 수도 있다. 편의를 위해 플라스틱 바가지도 여기저기 준비되어 있는데, 발을 씻는 행동은 다른 사람들과 교류하면서 발이 닿는 지면의 치유력을 공유하는 특별한 순간으로 전환된다.

이런 체험은 단순한 산 속 활동으로 보이지만, 자연과의 재연결 및 지역의 정체성 강화라는 심오한 의미를 내포한다. 콘크리트와 아스팔트가 지배하는 도시환경에서 땅은 현대적인 길 아래 숨겨져 있지만, 작은 산은 원시적인 자연 세계와 다시 연결시켜준다. 도시 생활의 제약에서 벗어나 맨발로 걷는 행위는 단순히 자연으로 돌아가는 것뿐 아니라, 우리가 공유하는 더 오래된 것으로의 회귀를 의미한다. 이로써 전통과 정체성은 과거 유물이 아닌, 집단적 자유로움과 땅과의 새로운 유대감 속에서 살아 있는 경험으로 자리 잡는다.

Legal Division of the Ground

Examining the invisible lines that legally divide the land allows for a deeper understanding of the uniqueness of these topographical features and their singular history, highlighting their value to the city.

This is due firstly to the geography and the particular shape of the land. The rugged terrain of these mini-mountains has undeniably influenced how the lots are divided. In contrast to flat land, where farmland or housing parcels typically define the landscape, the small and irregular lots on the mountains were divided based on the availability of land between steep slopes and natural features. The uneven ground dictated a meandering pattern of access routes and lot shapes.

Adding to this is the informal history of the hills' occupation. Although the original structures are no longer visible, the small, irregular lot divisions reveal the informal nature of their development. This is particularly evident in (12) 국사봉 *Guksabong*, where remnants of the irregular lot divisions from the former moon village remain. One of the rare examples still intact is (5) 응봉산 *Eungbongsan*, where the lots are irregular and shaped by the natural landscape, characterized by steep slopes. Here, the fragmented pattern of smaller lots follows the contours of the land, resulting in a distinctive neighbourhood of multi-level and narrow brick buildings that adapt to mountain's slopes.

토지의 법적 분할

토지를 법적으로 나누는 보이지 않는 분계선을 살펴보면, 지형적 특성과 독특한 역사를 더 깊게 이해하고 그곳이 도시에 어떤 가치를 제공하는지 알 수 있다.

우선 지형과 토지의 특정 형태 때문이다. 작은 산의 험준한 지형은 토지가 분할되는 방식에 확실히 영향을 미쳤다. 평지에서는 농지나 주거용 부지가 일반적으로 경관을 정하는 반면, 산에서는 가파른 경사와 자연적인 지형 사이의 토지 가용성에 따라 토지가 작고 불규칙하게 분할되었다. 고르지 않은 지면 때문에 접근 경로와 부지 모양이 구불구불한 패턴을 이룬다.

또한, 산등성을 점령했던 비공식적 역사가 영향을 미쳤다는 특징이 있다. 원래 구조는 더 이상 볼 수 없지만, 작고 불규칙한 토지 분할은 비공식적으로 개발된 흔적을 드러낸다. 특히 **국사봉 (12)** 에서 두드러지며, 이전 달동네의 불규칙한 토지 분할 흔적이 선명하다. 아직까지 온전한 흔적은 **응봉산 (5)** 에서 볼 수 있는데, 이곳 토지는 자연경관에 따라 불규칙하게 형성되었고 가파른 경사가 특징이다. 작게 분할된 토지 윤곽을 따라 산경사면에 지어진 좁은 벽돌 건물이 모인 독특한 거주지가 형성되어 있다.

Land use policy development during the recent development of Seoul reflects the shifting vision of urban development, as the government increasingly recognizes the value of natural assets within the city. It is worth highlighting that, despite the original uneven topography, the city could be more drastically covered with buildings without the establishment of legal protection of the land. Designated protected areas, such as Urban Nature Parks, Development Restricted Areas, Neighbourhood Parks, Natural Scenic Areas, Buffer Green Zones, and other park facilities, illustrate this changing perspective and are clearly represented on the land use maps. Today, a variety of mixed pattern land use can still be observed, as some mountains, such as (3) 봉화산 *Bonghwasan* or (24) 개화산 *Gaehwasan*, incorporate a patchwork of Urban Nature Park, Development Restricted Area and Neighbourhood Park.

Under the land use categories there persists an even greater discrepancy of lot division. For instance, in (12) 국사봉 *Guksabong*, several small lots remain, even though the area is officially designated as a Neighbourhood Park. This apparent inequality between private and public use of the ground reflects the slow and often controversial process of land acquisition. In contrast, (20) 우장산 *Ujangsan* displays a more resolved division of land, with full alignment between the district's development plans and the mountain's use as a Neighbourhood Park.

최근 서울의 토지이용 정책 개발은 정부가 도시 속 자연 자산의 가치를 더욱 인정하기 시작하면서 변화한 도시 개발 비전을 보여준다. 눈여겨볼 점은, 법적 보호 장치가 없었다면 지형이 고르지 않음에도 서울은 훨씬 더 무분별하게 건물로 덮일 수 있었다는 사실이다. 도시 자연공원, 개발제한구역, 근린공원, 자연경관 구역, 완충녹지, 기타 공원 시설 등 지정된 보호구역은 이런 변화된 관점을 반영하며, 토지 용도 지도에도 확실히 표시된다. 요즘도 혼합 용도 토지의 다양한 사용을 볼 수 있는데, **봉화산** (3) 또는 **개화산** (24) 같은 일부 산에는 도시 자연공원, 개발제한구역, 근린공원이 모두 있다.

토지 용도 범주에서는 이런 혼합 용도 토지가 토지 분할을 더 복잡한 과정으로 만든다. 예를 들어 **국사봉** (12) 은 근린공원으로 공식 지정되었지만, 일부 작은 토지가 그대로 남아 있다. 사적 용도와 공공 용도 사이의 이런 불균형이 발생하는 것은 토지 취득 과정이 느리고 종종 논란의 여지가 있기 때문이다. 대조적으로 **우장산** (20) 은 지방자치단체의 개발계획과 산을 근린공원으로 사용하려는 용도가 완전히 일치해서 토지 분할이 확실하다.

(1) 천장산 *Cheonjangsan*

(5) 응봉산 *Eungbongsan*

③ 봉화산 Bonghwasan
⑦ 궁동 Gungdong
⑥ 와우산 Wausan
⑳ 우장산 Ujangsan
Cadastral Outline
Lot Division Lines
Urban Nature Park
Development Restricted Area
Natural Environment Conservation Area
Neighbourhood Park
Natural Scenic Area
Other Park Facilities
Buffer Green Area
Agriculture and Forestry Area

Looking at land use also offers a critical lens for understanding the current status of the mountains and envisioning their future. Without clear legal protections to shield the mountains from urban development, these essential city assets will remain at risk. Supporting a vision of the city where natural buffers punctuate the urban landscape is vital if they are to be preserved. The historical and ecological significance of these areas cannot be understated, particularly their role in climate regulation, biodiversity preservation, and maintaining a habitat for wildlife. Given the scale and significance of Seoul's megapolis, these territories offering a balance between human and non-human inhabitants of the city which is both unique and valuable.

Establishing and maintaining clear legal boundaries is, therefore, essential to resist development pressures. However, certain areas still have ambiguous lot divisions and land use classifications, leaving them vulnerable. Clarifying and reinforcing the legal framework that defines unbuildable territory could offer stronger protection for these spaces. In fact, this approach could prove more effective than the current trend of creating "green corridors" within the city, which often require clearing existing built areas. Instead of following fashionable urban development strategies, could the city place greater value on preserving what already exists?

앞에 표시된 토지 용도는 산의 현재 상태를 이해하고 미래를 구상하는 데 중요한 단서를 제공한다. 도시 개발에서 산을 보호하는 명확한 법률이 없으면 도시의 필수 자산인 산은 위험에 처한다. 완충녹지를 보호하기 위해서는 자연적인 완충지가 도시경관을 이루는 도시계획을 지원해야 한다. 이런 지역은 역사적, 생태적 측면에서 매우 중요한 가치를 지니며, 특히 기후 규제, 생물 다양성 보전, 야생 동식물 서식지 보존에 중요하다. 거대도시인 서울의 규모와 중요성을 고려할 때 이런 완충녹지는 인간과 동식물 사이에 유의미한 균형을 제공한다.

따라서 명확한 법적 경계를 설정하고 유지하는 것은 개발 압력에 저항하는 데 필수적이다. 하지만 여전히 일부 지역은 토지 분할과 용도 분류가 모호해서 취약한 상황에 처해 있다. 건설 제한 구역을 정의하는 법적 틀을 명확히 정하면 녹지를 더 강력하게 보호할 수 있다. 이런 접근법은 기존 건축물을 재정비해서 도시 속 녹지를 조성하는 현재 추세보다 훨씬 더 효과적일 수 있다. 유행하는 도시 개발 전략을 따르는 대신, 도시 속 이미 존재하는 자연 자산을 보존하는 일을 더 중요시할 수는 없을까?

Reinvesting in Seoul's mountains requires a careful evaluation of the unique qualities of each site. It is important to maintain a diversity of scenarios within mountain territories—balancing areas for public enjoyment with those that offer limited accessibility. The goal should be to avoid turning the mountains into consumable land saturated with activities. While some areas would benefit from public facilities and infrastructure improvements, other zones should be left untouched, allowing a deeper level of wilderness to flourish within the city.

Finally, the valuable engagement of local residents, such as in ③ 봉화산 *Bonghwasan*, should be appreciated—even if the land is legally classified as public. Maintaining a gradient of personal involvement and private initiatives in the care of the land adds another layer of connection between the community and the landscape. The mountains should be seen as shared assets, fully accessible to everyone. However, allowing a personal, and sometimes semi-private, connection with the hills—especially in terms of maintenance—could transform residents from mere consumers into caretakers of these spaces.

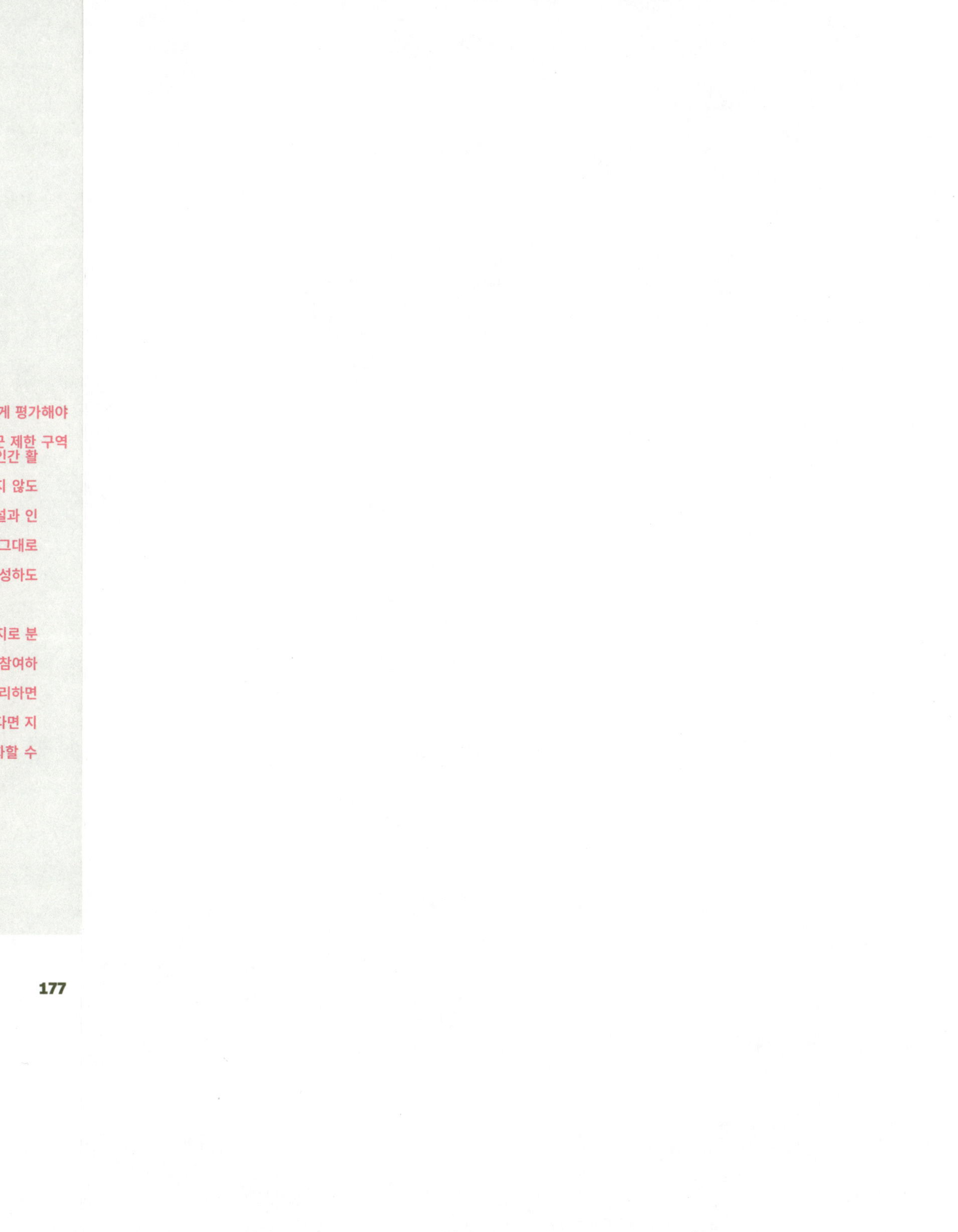

서울의 산에 재투자하기 위해서는 각 지역의 독특한 특성을 신중하게 평가해야 한다. 산 부지의 다양한 활동을 지속함으로써 공공 여가 구역과 접근 제한 구역 사이의 균형을 맞추는 것이 중요하다. 또한 인간 활동으로 가득 차서 산이 소모적인 용지가 되지 않도록 목표를 정해야 한다. 일부 지역은 공공시설과 인프라 개선의 혜택을 받겠지만, 다른 지역은 그대로 보존하여 도심에서도 야생 동식물이 더욱 번성하도록 해야 한다.

마지막으로, 용도가 법적으로 공공 토지로 분류되더라도 **봉화산** ③처럼 지역 주민이 참여하는 값진 활동에 감사해야 한다. 토지를 잘 관리하면서도 개인의 자발적인 참여와 노력이 지속된다면 지역 커뮤니티와 자연경관의 연결을 더욱 강화할 수 있다. 산은 모두가 온전히 접근할 수 있는 공유 자산으로 보아야 한다. 하지만 주민들이 산과 개인적인 관계를 맺도록 허용한다면, 특히 유지 보수 차원에서 그들은 단순한 소비자를 넘어 산을 진정으로 돌보는 관리자로 거듭날 수 있을 것이다.

찾아보기

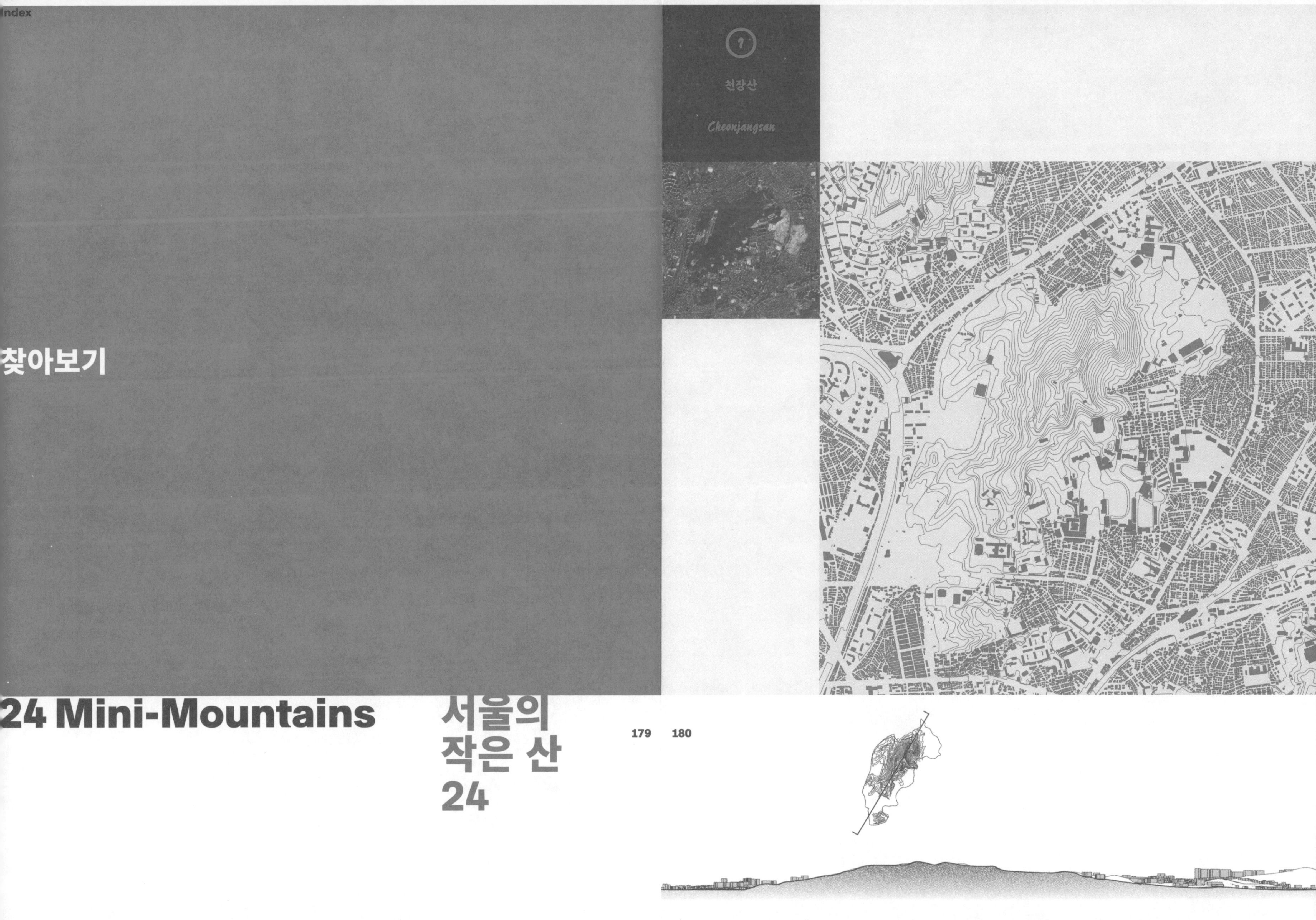

24 Mini-Mountains

서울의
작은 산
24

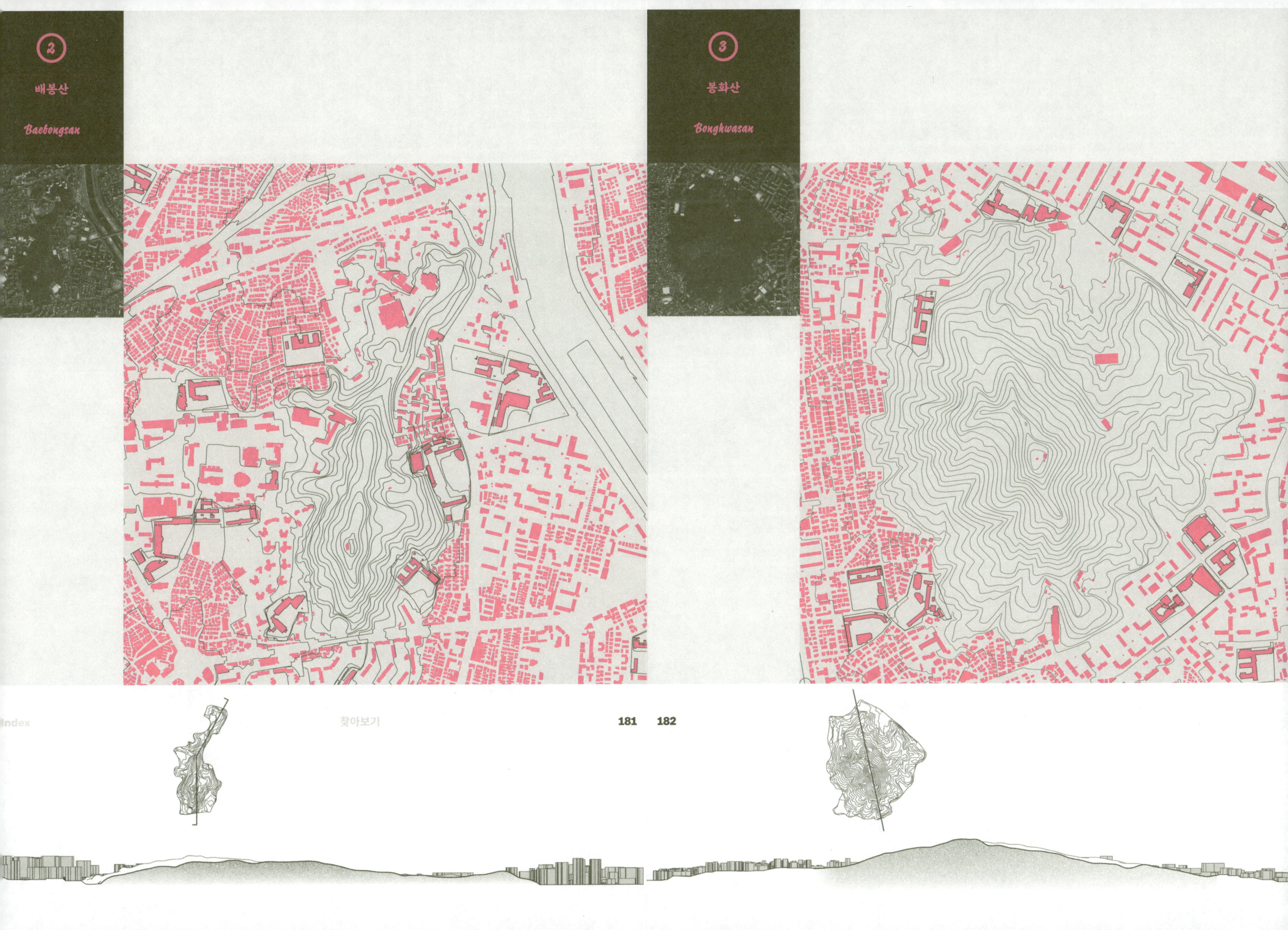
2
배봉산
Baebongsan
3
봉화산
Bonghwasan

4
영축산
Yeongchuksan
5
응봉산
Eungbongsan

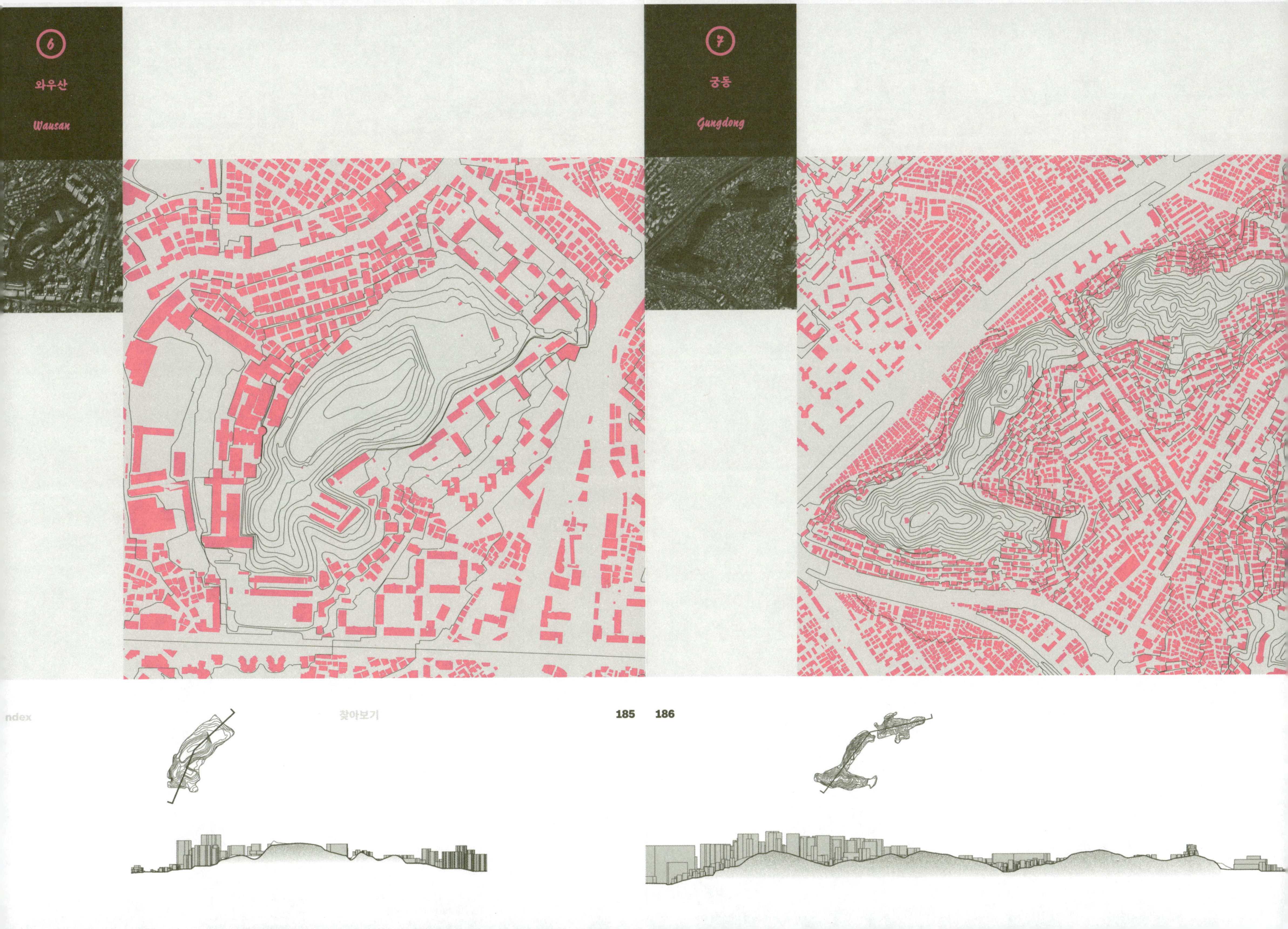
6
와우산
Wausan
7
궁동
Gungdong

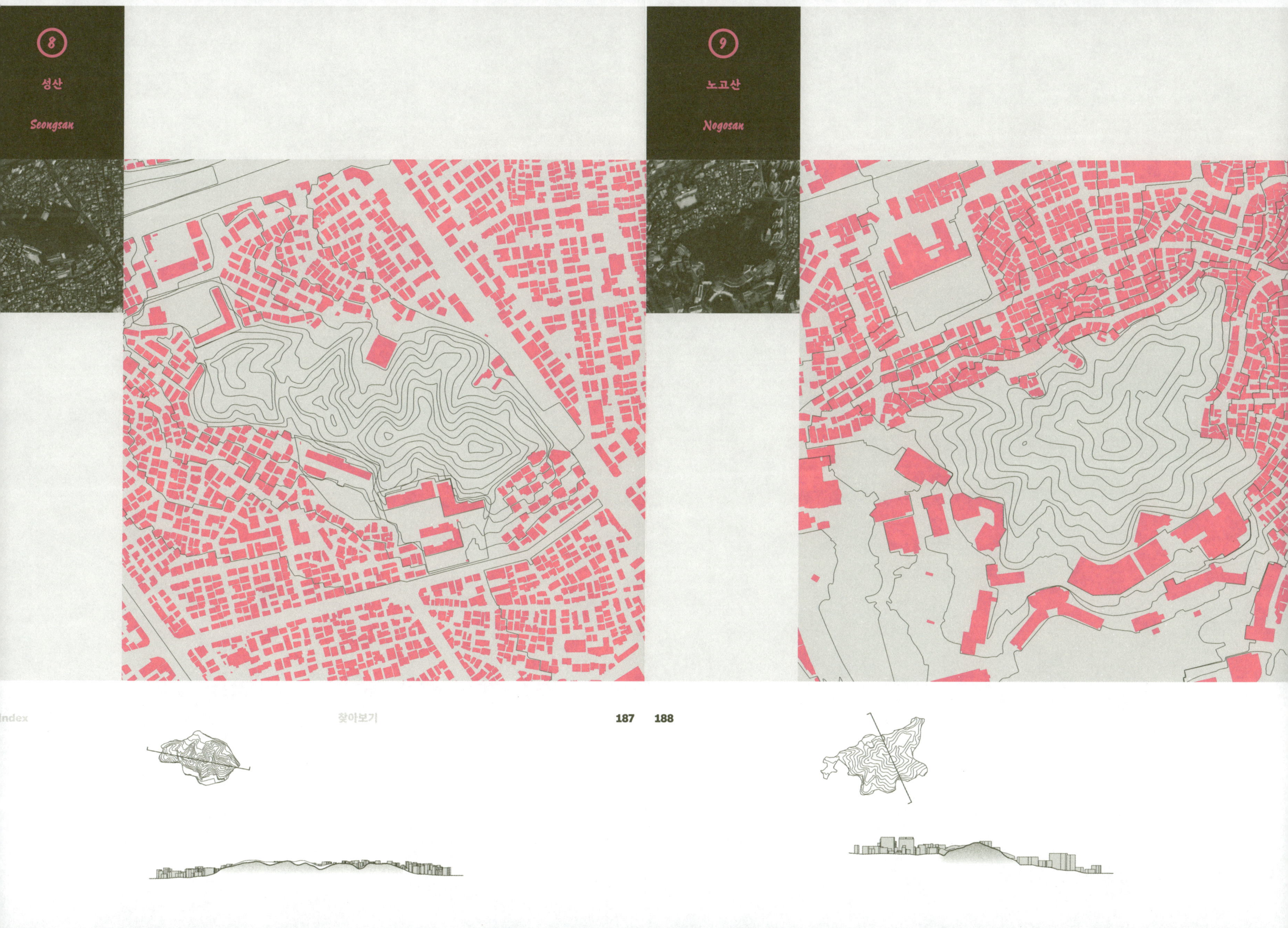
8
성산
Seongsan
9
노고산
Nogosan

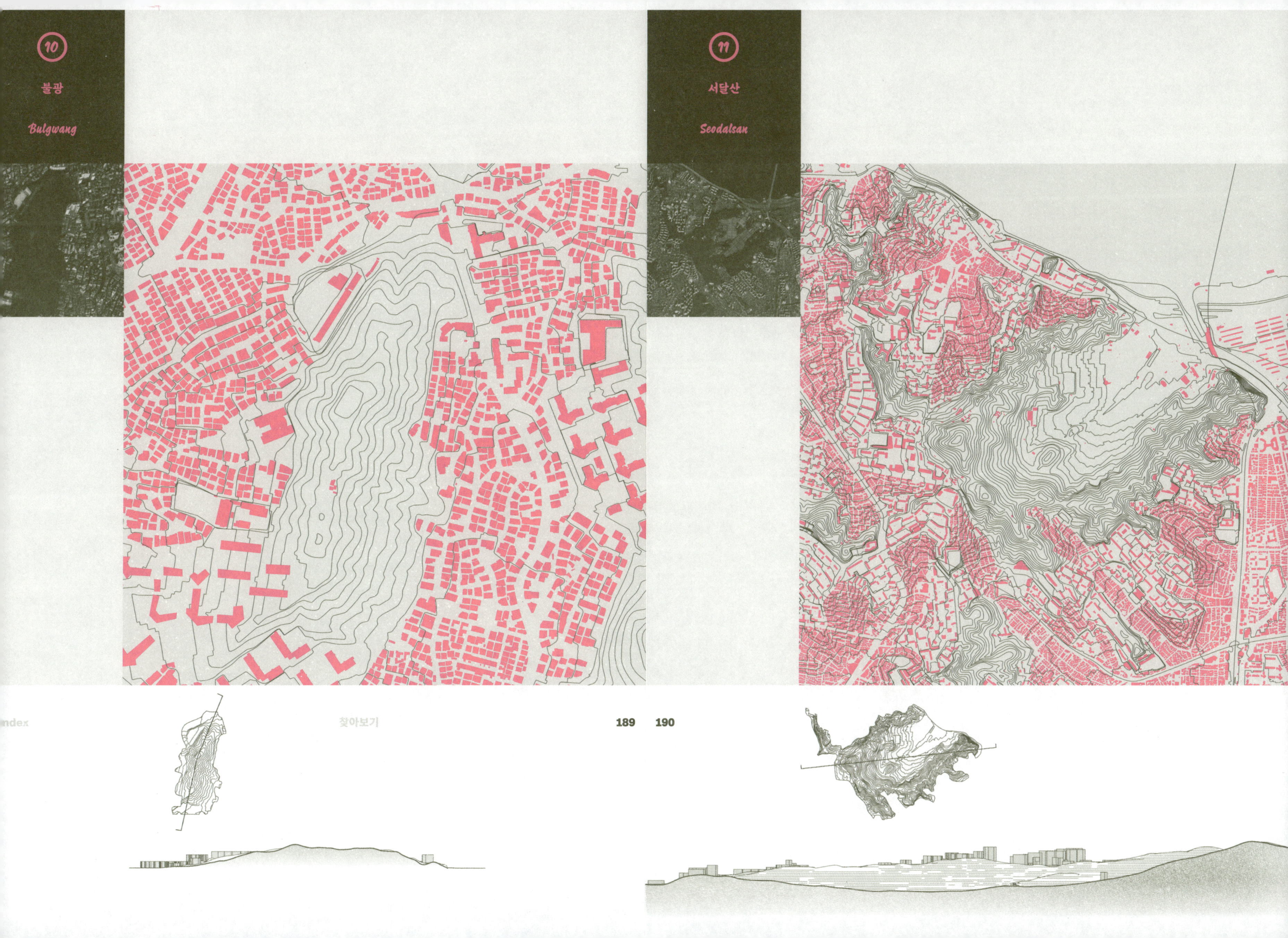
10
불광
Bulgwang
11
서달산
Seodalsan

12
국사봉
Guksabong
13
용마산
Yongmasan

14
장군봉
Janggunbong
15
까치산
Kkachisan

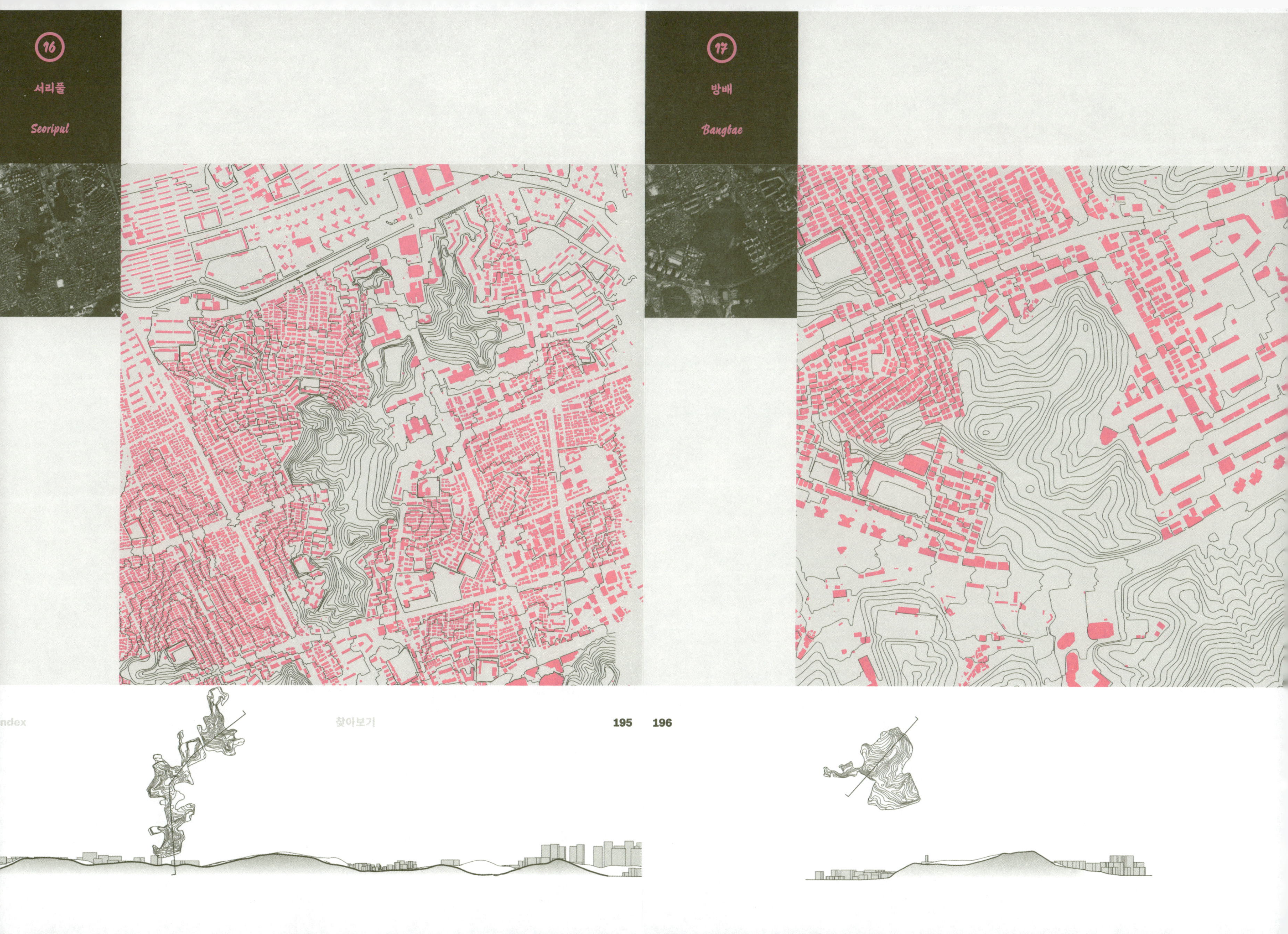
16
서리풀
Seoripul
17
방배
Bangbae

18
매봉산
Maebongsan
19
봉제산
Bongjesan

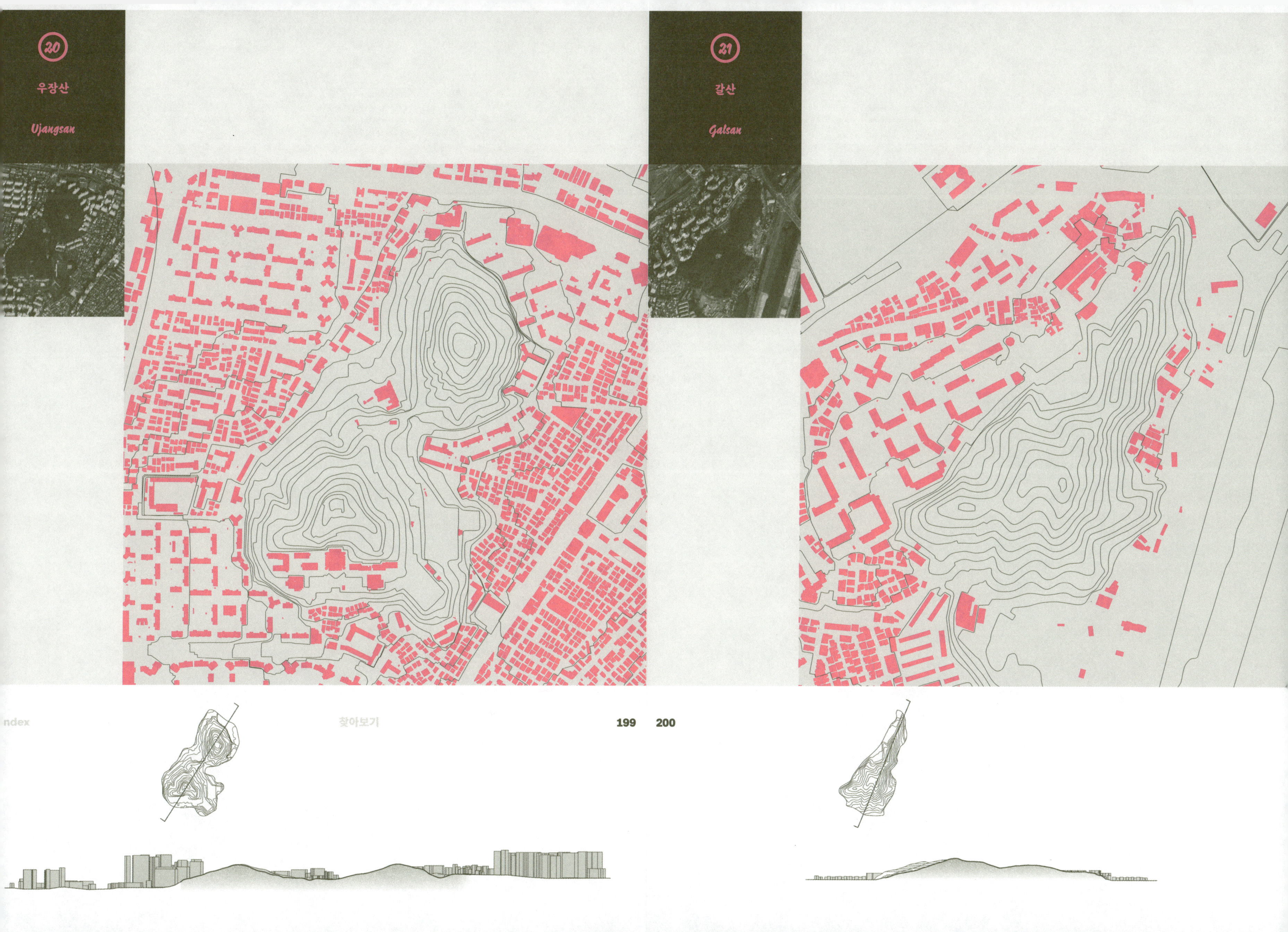
20
우장산
Ujangsan
21
갈산
Galsan

22
계남
Gyenam
23
개웅산
Gaeungsan

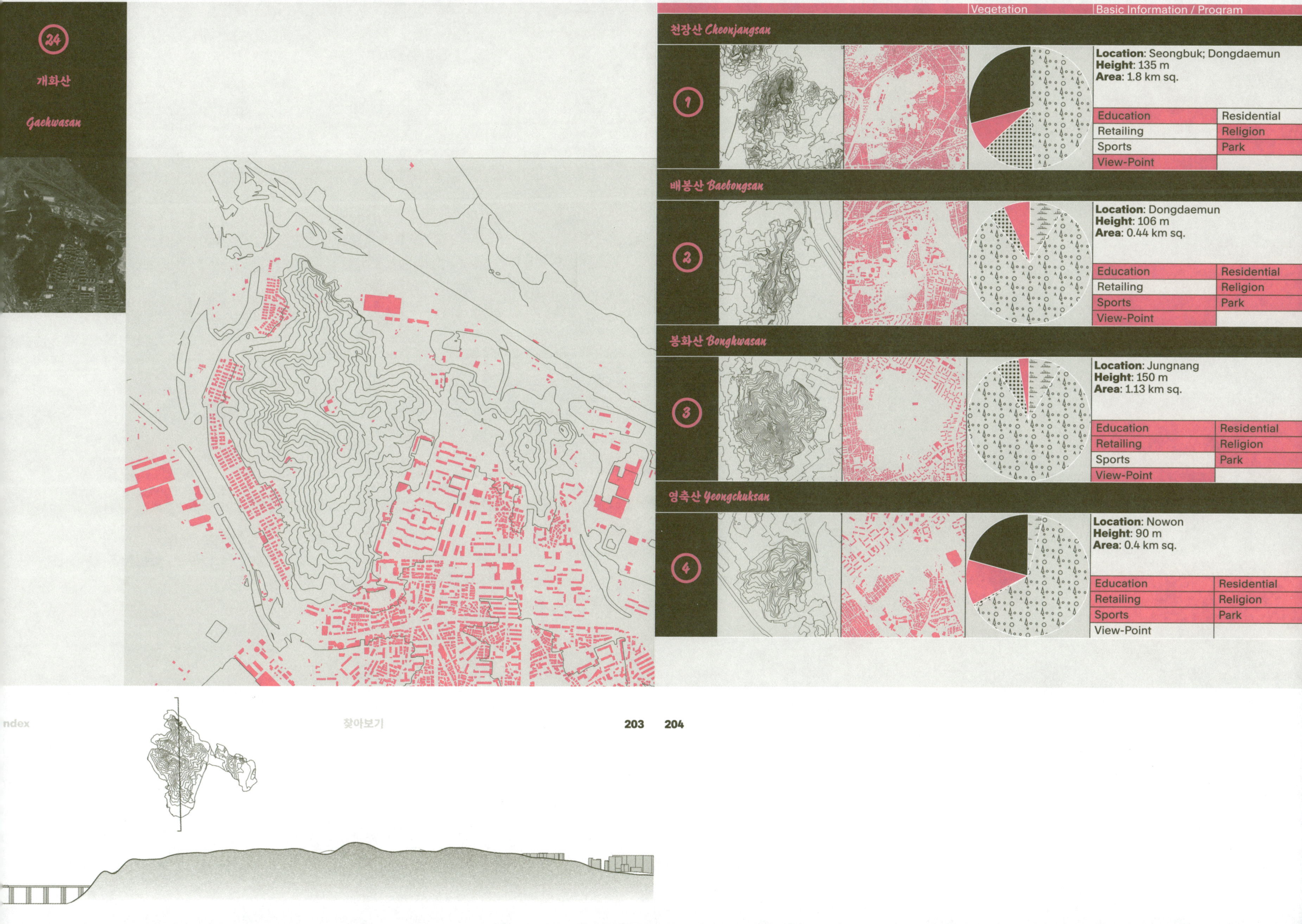
24
개화산
Gaehwasan
Vegetation
Basic Information / Program
천장산 Cheonjangsan
1
Location: Seongbuk; Dongdaemun
Height: 135 m
Area: 1.8 km sq.
Education
Residential
Retailing
Religion
Sports
Park
View-Point
배봉산 Baebongsan
2
Location: Dongdaemun
Height: 106 m
Area: 0.44 km sq.
Education
Residential
Retailing
Religion
Sports
Park
View-Point
봉화산 Bonghwasan
3
Location: Jungnang
Height: 150 m
Area: 1.13 km sq.
Education
Residential
Retailing
Religion
Sports
Park
View-Point
영축산 Yeongchuksan
4
Location: Nowon
Height: 90 m
Area: 0.4 km sq.
Education
Residential
Retailing
Religion
Sports
Park
View-Point

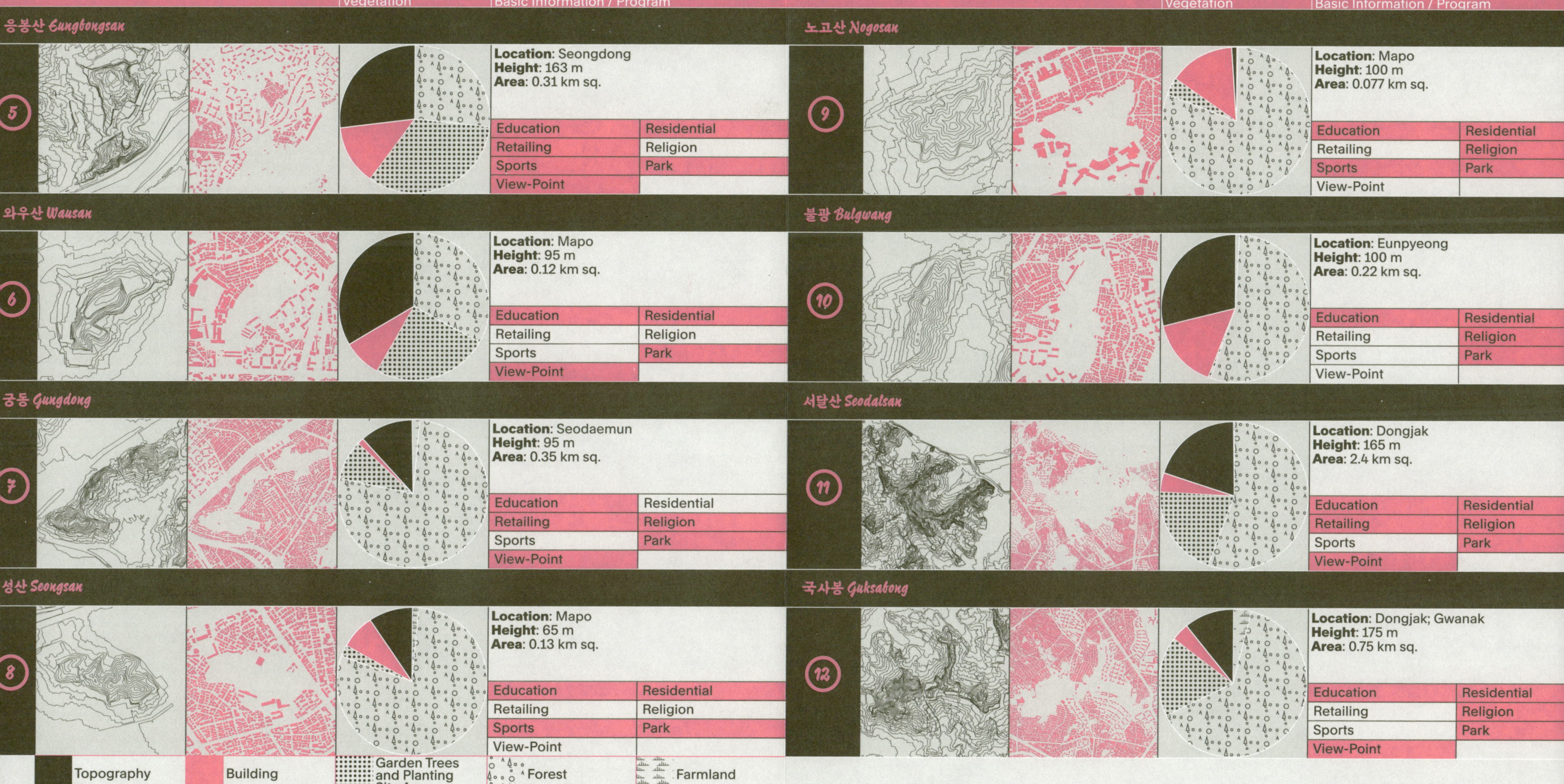
Vegetation
Basic Information / Program
응봉산 Eungbongsan
5
Location: Seongdong
Height: 163 m
Area: 0.31 km sq.
Education
Residential
Retailing
Religion
Sports
Park
View-Point
와우산 Wausan
6
Location: Mapo
Height: 95 m
Area: 0.12 km sq.
Education
Residential
Retailing
Religion
Sports
Park
View-Point
궁동 Gungdong
7
Location: Seodaemun
Height: 95 m
Area: 0.35 km sq.
Education
Residential
Retailing
Religion
Sports
Park
View-Point
성산 Seongsan
8
Location: Mapo
Height: 65 m
Area: 0.13 km sq.
Education
Residential
Retailing
Religion
Sports
Park
View-Point
Topography
Building
Garden Trees and Planting Site Area
Forest
Farmland
Vegetation
Basic Information / Program
노고산 Nogosan
9
Location: Mapo
Height: 100 m
Area: 0.077 km sq.
Education
Residential
Retailing
Religion
Sports
Park
View-Point
불광 Bulgwang
10
Location: Eunpyeong
Height: 100 m
Area: 0.22 km sq.
Education
Residential
Retailing
Religion
Sports
Park
View-Point
서달산 Seodalsan
11
Location: Dongjak
Height: 165 m
Area: 2.4 km sq.
Education
Residential
Retailing
Religion
Sports
Park
View-Point
국사봉 Guksabong
12
Location: Dongjak; Gwanak
Height: 175 m
Area: 0.75 km sq.
Education
Residential
Retailing
Religion
Sports
Park
View-Point

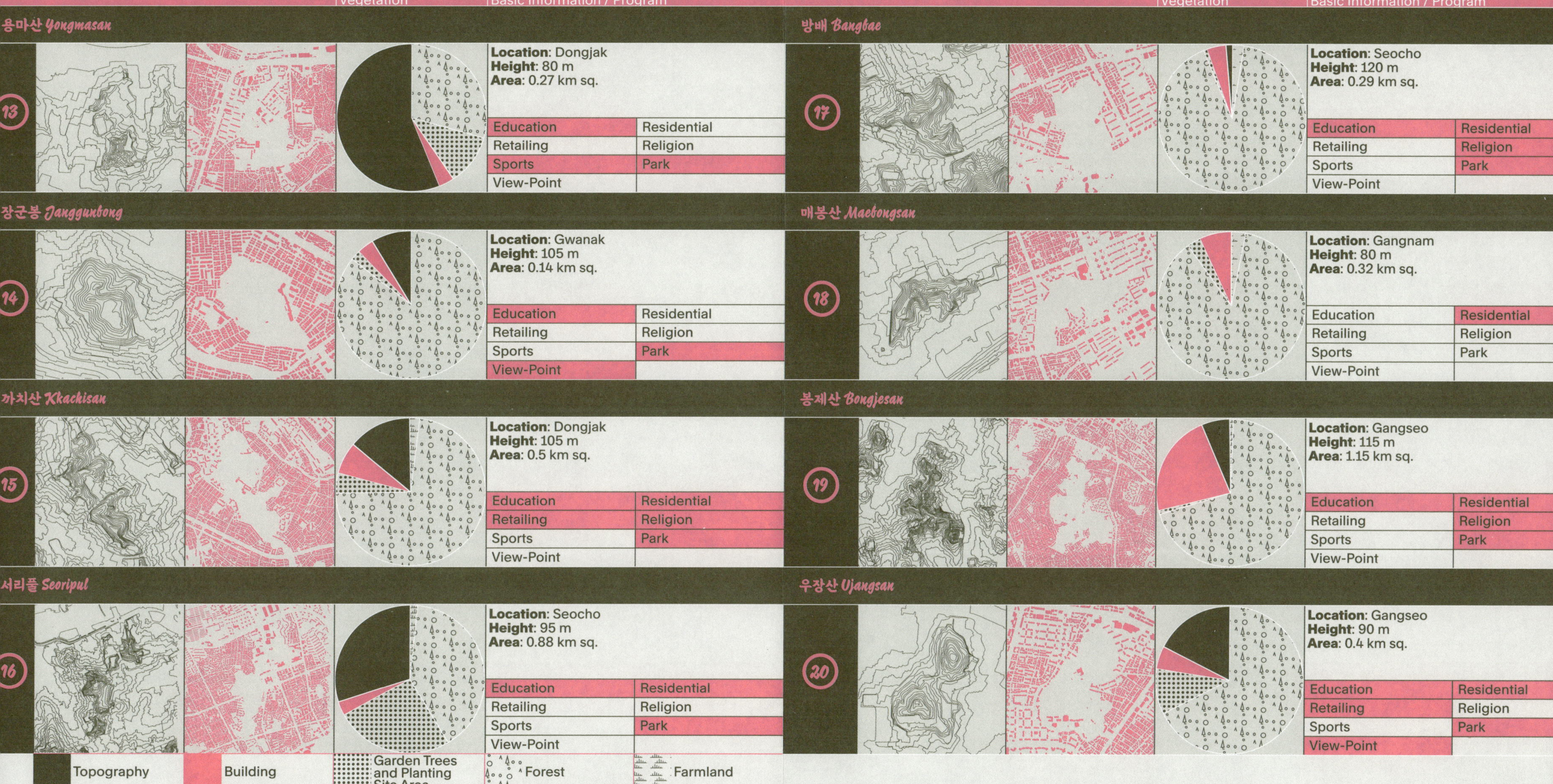

Vegetation
Basic Information / Program
용마산 Yongmasan
13
Location: Dongjak
Height: 80 m
Area: 0.27 km sq.
Education
Residential
Retailing
Religion
Sports
Park
View-Point
장군봉 Janggunbong
14
Location: Gwanak
Height: 105 m
Area: 0.14 km sq.
Education
Residential
Retailing
Religion
Sports
Park
View-Point
까치산 Kkachisan
15
Location: Dongjak
Height: 105 m
Area: 0.5 km sq.
Education
Residential
Retailing
Religion
Sports
Park
View-Point
서리풀 Seoripul
16
Location: Seocho
Height: 95 m
Area: 0.88 km sq.
Education
Residential
Retailing
Religion
Sports
Park
View-Point
Topography
Building
Garden Trees and Planting Site Area
Forest
Farmland
Vegetation
Basic Information / Program
방배 Bangbae
17
Location: Seocho
Height: 120 m
Area: 0.29 km sq.
Education
Residential
Retailing
Religion
Sports
Park
View-Point
매봉산 Maebongsan
18
Location: Gangnam
Height: 80 m
Area: 0.32 km sq.
Education
Residential
Retailing
Religion
Sports
Park
View-Point
봉제산 Bongjesan
19
Location: Gangseo
Height: 115 m
Area: 1.15 km sq.
Education
Residential
Retailing
Religion
Sports
Park
View-Point
우장산 Ujangsan
20
Location: Gangseo
Height: 90 m
Area: 0.4 km sq.
Education
Residential
Retailing
Religion
Sports
Park
View-Point

Vegetation
Basic Information / Program
갈산 Galsan
21
Location: Yangcheon
Height: 70 m
Area: 0.14 km sq.
Education
Residential
Retailing
Religion
Sports
Park
View-Point
계남 Gyenam
22
Location: Yangcheon
Height: 75 m
Area: 0.73 km sq.
Education
Residential
Retailing
Religion
Sports
Park
View-Point
개웅산 Gaeungsan
23
Location: Guro
Height: 110 m
Area: 0.71 km sq.
Education
Residential
Retailing
Religion
Sports
Park
View-Point
개화산 Gaehwasan
24
Location: Gangseo
Height: 132 m
Area: 1.83 km sq.
Education
Residential
Retailing
Religion
Sports
Park
View-Point
Topography
Building
Garden Trees and Planting Site Area
Forest
Farmland

The research project, *Seoul Mini-Mountains: Reading the City from its Topographical Gaps*, was awarded funding from the General Research Fund (GRF), granted by the University Grants Committee (UGC) in Hong Kong in 2023. Initiated in 2019, the project began with the discovery of small, unexpected topographical features in Seoul, found while traversing the city on foot. Over the years, the project evolved through off-site and on-site observations, and the creation of a comprehensive series of analytical drawings, aiming to capture the intricate processes behind the formation and transformation of these unique spaces. This work has inspired and enriched three courses at the Department of Architecture, University of Hong Kong, fostering an academic exploration of Seoul's urban voids. The research in progress was presented at the Global Studio Exhibition during the 2019 Seoul Biennale of Architecture and Urbanism and at the Seoul Hall of Urbanism and Architecture in 2023. It culminates in an exhibition at the Seoul Museum of History in winter 2024–25, inviting reflection on the often-overlooked spaces within our cities and the stories they tell.

서울의 작은 산: 보이드를 통해 도시를 읽다’ 프로젝트는 2023년 홍콩대학교육위원회(UGC)의 일반 연구 기금(GRF)을 지원받아 진행되었다. 이 연구는 2019년에 시작되었으며, 그 시작은 서울을 걸으며 발견한 작은 단위의 예상치 못한 지형에서 비롯되었다. 수년 동안 현장 관찰과 자료 수집, 포괄적인 분석 그림을 제작하면서 연구를 심화해 나갔고, 이 독특한 공간들이 형성되고 변화하는 복잡한 과정을 밝히는 데 초점을 맞췄다. 이 연구는 홍콩대학교 건축학과의 세 개 교과과정에 영감을 주었고, 서울의 보이드 공간에 대한 학문적 탐구를 심화시키는 계기가 되었다. 현재까지 진행 중인 연구는 2019년 서울도시건축비엔날레 글로벌 스튜디오 전시 및 2023년 서울 도시건축 전시관에서 소개되었으며, 2024-2025년 겨울에는 서울역사박물관에서 전시되었다. 이 프로젝트는 우리가 자주 간과하는 도시 공간과 그 공간의 배경 이야기를 다시 생각해 보는 기회를 제공할 것이다.

② 배봉산 Baebongsan

⑥ 와우산 *Wausan*

⑦ 천장산 *Cheonjangsan*

12 국사봉 *Guksabong*

Image Sources

Cover
표지
New Edition of the Restricted Development Zone of Seoul, 1974, source: Seoul Museum of History.
개정서울특별시전도 (改定서울特別市全圖), 1974년, 출처: 서울역사박물관.

Chapter 1: History
1장: 역사

p. 50, fig. 1:
Fengshui Guidebook 지리현주, mid-Joseon Dynasty, source: National Folk Museum of Korea, Seoul.
풍수 가이드북 〈지리현주〉, 조선 중기, 출처: 국립민속박물관, 서울.

p. 52, fig. 3:
Panoramic View of Gyeongseong 京城全景, Seoul, 1904, source: Seoul Museum of History.
경성의 파노라마 전경, 서울, 1904년, 출처: 서울역사박물관.

p. 58, fig. 1:
Hanyangdoseongdo 한성도, Seoul, 1770, source: Leeum, Samsung Museum of Art, Seoul.
한양도성도, 서울, 1770년, 출처: 리움(Leeum) 삼성미술관.

p. 60, fig. 1:
Sasangeumpyodo 사산금표도, Seoul, 1765, source: Seoul Museum of History.
사산금표도, 서울, 1765년, 출처: 서울역사박물관

p. 60, fig. 2:
Gyeongjoobudo 경조오부도, 1861, source: Seoul Museum of History.
경조오부도, 1861년, 출처: 서울역사박물관.

p. 62, fig. 1:
Gyeongseongdo 경성도, Seoul, 1922, source: Seoul Museum of History, Seoul.
경성도, 서울, 1922년, 출처: 서울역사박물관.

p. 64, fig. 1:
Kyongsong or Seoul (Keijo), 1946, United States, Army Map Service, source: University of Wisconsin, The American Geographical Society Library Digital Map Collection.
경성 또는 서울(게이조), 1946년, 미국 육군 지도 서비스, 출처: 위스콘신 대학교, 미국 지리학회 도서관 디지털 지도 컬렉션.

p. 66, fig. 1:
New Edition of the Restricted Development Zone of Seoul, 1975, source: Seoul Museum of History.
서울 개발 제한 구역 개정판, 1975년, 출처: 서울역사박물관.

p. 84, fig. 1:
Gaehwasa Temple 개화사, 1740–1750, author: Jeong Seon 정선, source: Kangsong Art Museum, Seoul.
개화사, 1740–1750, 작가: 정선, 출처: 간송미술관, 서울.

p. 84, fig. 2:
Map in Yangcheon Eupji, 1899, source: Digital collection Jangseogak.
양천읍지 지도, 1899년, 출처: 장서각 디지털 컬렉션.

p. 86, fig. 2:
Manguri Cemetery, Seoul, 1996, source: The Seoul Research Data Service.
망우리 공동묘지, 서울, 1996년, 출처: 서울리서치 데이터 서비스.

Chapter 2: Boundaries
2장: 경계

p. 96, fig. 1:
Demolition of unauthorized buildings in Hannam-dong in 1962, 1962, source: Major City Photo Collection, Seoul Record Center.
1962년 한남동의 불법 건물 철거, 1962년, 출처: 서울기록관 주요 도시 사진 컬렉션.

p. 102, fig. 1:
Panoramic View of Cheongyeon-dong Shantytown before building Geumhwa Apartment, Cheongyeon-dong, Seoul, 1968, source: Seoul Museum of History.
서울 청연동 금화아파트 건설 전에 청연동 판자촌의 파노라마 전경, 1968년, 출처: 서울역사박물관.

p. 108, figs. 1–3:
Construction Site of Geumhwa Citizen Apartment (1), Yongsan District, Seoul, 1968, source: Seoul Museum of History.
서울 용산구 금화시민아파트 건설 현장(1), 1968년, 출처: 서울역사박물관.

fig. 4:
Geumhwa Apartment, Geumhwa-dong, Seoul, 1971, source: Seoul Museum of History.
서울 금화동의 금화아파트, 1971년, 출처: 서울역사박물관.

p. 110, fig. 2:
The Rescue Operation Site of the Apartment 1970.04.10., source: Seoul Museum of History.
아파트 구조작업 현장 1970.04.10, 출처: 서울역사박물관.

p. 112, fig. 1:
Seoul Panorama View from Ansan, photo: Inho Choi, 2009, source: Seoul Museum of History.
안산에서 본 서울 파노라마, 사진: 최인호, 2009년, 출처: 서울역사박물관.

Chapter 3: Activities
3장: 경험

pp. 132
Park-planning Map of Seoul, 1920s, source: Seoul Museum of History.
서울공원계획지도, 1920년대, 출처: 서울역사박물관.

pp. 134
Seoul Metropolitan City Urban Planning Park Change Before and After Plan (Seoul Metropolitan City Planning Park Change Plan),1950s., source: Seoul Museum of History.
서울특별시 도시계획공원 변경 전후 계획도(서울특별시), 1950년대, 출처: 서울역사박물관.

pp. 136
New Edition of the Restricted Development Zone of Seoul, 1974, source: Seoul Museum of History.
개정서울특별시전도 (改定서울特別市全圖), 1974년, 출처: 서울역사박물관.

pp. 142
Namsan Erosion Control Construction Site, 1962, source: Seoul Museum of History.
남산방재공사 현장, 1962년, 출처: 서울역사박물관.

Bibliography

Bengston, David N., and Yeo-Chang Youn. "Urban Containment Policies and the Protection of Natural Areas: The Case of Seoul's Greenbelt." *Ecology and Society* 11, no. 1 (2006). http://www.jstor.org/stable/26267777.

Borio, G. *Looking for the Voids: Learning from Asia's Liminal Urban Spaces as a Foundation to Expand an Architectural Practice.* Zurich: Park Books, 2023.

Borio, G., and C. Wüthrich. *Hong Kong In-Between.* Hong Kong: MCCM Creations; Zurich: Park Books, 2015.

Borio, G., and C. Wüthrich. *The People of Duckling Hill.* Hong Kong: Parallel Lab & The Hong Kong Polytechnic University, 2016.

Cho, Jeon-hwan. "Courtyards of Traditional Korean Houses." *Koreana* 27 (Autumn 2013): 18-27.

Choi, C. J. "P'ungsu, the Korean Traditional Geographic Thoughts." *Korea Journal* 26, no. 5 (1986): 35–45.

Choi, H. S. "New Forms of Place-Making and Public Space in Contemporary Urban Development in Seoul." *Current Urban Studies* 5, no. 3 (2017): 332–47.

Choi, Jae-Soon. *Hanoak: Traditional Korean Homes.* Elizabeth, NJ: Hollym, 1999.

Choi, J. H. "A Study on Constructional Intention, Idea, Thought and Aesthetic Consciousness of Joseon Royal Tombs." *Journal of Korean Institute of Traditional Landscape Architecture* 34, no. 4 (December 19, 2016): 66–77.

Chung, M. S. *Encyclopedia of Korean Folk Beliefs.* Seoul: The National Folk Museum of Korea, 2013.

Encyclopedia of Korean Culture. "산신당(山神堂)." Accessed April 28, 2024. https://encykorea.aks.ac.kr/Article/E0026268.

Ginzburg, C., and C. Poni. *La Micro-Histoire.* Paris: Gallimard-Le Débat, 1981.

Grayson, J. H. "Female Mountain Spirits in Korea: A Neglected Tradition." *Asian Folklore Studies* 55 (1996): 119–34.

Habraken, J. *The Structure of the Ordinary: Form and Control in the Built Environment.* Cambridge, MA: The MIT Press, 2000.

Jackson, Ben. *Korean Architecture: Breathing with Nature.* Korea Essentials, no. 12. Seoul: Seoul Selection, 2012.

Jang, Yong Hae, and Lee, Young Han. "The Study on Street Facade Characteristics of Junggye-dong 104 Village in Seoul." *Journal of the Korea Institute of Ecological Architecture and Environment* 13, no. 5 (October 2013): 5–15. https://doi.org/10.12813/kieae.2013.13.5.005.

Jee, E. A. "Cities on the Edge: Significance and Preservation of Hillside Squatter Settlements in Korea." Master's thesis, Columbia University, 2014.

Jo, Sanku. "Korean House Hanok." Accessed October 7, 2024. https://artsandculture.google.com/exhibit/korean-house-hanok-jo-sanku/gQR542M5?hl=en.

Johnston, R, and Cybriwsky, R,. "Spatial Behavior: A Geographic Perspective, by Reginald G. Golledge and Robert J. Stimson; *Seoul: The Making of a Metropolis*, by Joochul Kim and Sang-Chuel Choe." *Urban Geography* 19, no. 6 (1998): 582–584. https://doi.org/10.2747/0272-3638.19.6.582.

Kim, H. M., and Sun Sheng Han. "Seoul." *Cities* 29, no. 2 (2012): 142–54. https://doi.org/10.1016/j.cities.2011.02.003.

Kim, S-E T. "The Re-Emergence of Chosŏn Buddhism in the 17th Century: A Question of Institutional Development and Legitimation." *Journal of Korean Religions* 10, no. 2 (2019): 221–46. https://doi.org/10.1353/jkr.2019.0011.

Kim, S-W. *Urban Planning & Management.* Seoul: Seoul Institute, Policy Area: Urban Planning.

Kim, K-J, ed. *Seoul Twentieth Century: Growth & Change of the Last 100 Years.* Seoul: Seoul Development Institute, 2003.

Kim, M-S, Lee K-J, Kim J-Y, and Hur J-Y. "A Study on the Change and Management of Historical Landscape Forest of Taeneung, Joseon Dynasty Royal Tomb, Seoul, Korea." *Journal of the Korean Institute of Landscape Architecture* 43, no. 2 (April 2015): 56–72. https://doi.org/10.9715/KILA.2015.43.2.056.

Kim, N-C,. "Ecological Restoration and Revegetation Works in Korea." *Landscape and Ecological Engineering* 1 (2005): 77–83. https://doi.org/10.1007/s11355-005-0011-3.

Korea Forest Service. "Korean Forests at a Glance." Accessed October 7, 2024. https://english.forest.go.kr

Lee, D. K., and Hyeyeong Choe. "Estimating the Impacts of Urban Expansion on Landscape Ecology: Forestland Perspective in the Greater Seoul Metropolitan Area." *Journal of Urban Planning and Development* 137, no. 4 (2011): 425–37. https://doi.org/10.1061/(ASCE)UP.1943-5444.0000090.

Lee, M. R., ed. *The Statistical Yearbook of Forestry 2023 (No.53).* Forest Statistical System, 2023. https://kfss.forest.go.kr/stat/ptl/fyb/frstyYrBookList.do?curMenu=9854.

Lee, S-c, and H- Park. *Study on Landscape Management Evaluation and Improvement Plans according to Seoul Landscape Plan.* Seoul: Seoul Institute, 2011.

Lee, S. H. "Continuity and Consistency of the Traditional Courtyard House Plan in Modern Korean Dwellings." *Traditional Dwellings and Settlements Review* 3, no. 1 (1991): 65–76.

Maeng, D-M. *Residential Environment Improvement Program f or the City of Seoul.* Seoul: Seoul Institute, Policy Area: Housing.

Maki, F., et al. *City With A Hidden Past.* Tokyo: Kajima Institute, 2018.

Mason, D. A. *Spirit of the Mountains.* Elizabeth, NJ: Hollym International Corp., 1999.

Mobrand, E. "Struggles over Unlicensed Housing in Seoul, 1960-80." *Urban Studies* 45, no. 2 (2008): 367–89. https://doi.org/10.1177/0042098007085968.

Nam, J. H., and Heungsoon Kim. "Studies on Usage Patterns and Use Range of Neighborhood Parks: Focused on 'Regional Area Parks' in Seoul, Korea." *Journal of Asian Architecture and Building Engineering* 15, no. 3 (2016): 495–501. https://doi.org/10.3130/jaabe.15.495.

Park, H. C. *Landscape Management Policy for Better Seoul.* Seoul: Seoul Metropolitan Government and The Seoul Institute, 2017.

Seoul Development Institute. *History of Housing Redevelopment of Seoul.* 1996.

Seoul Metropolitan City. *Namsan Recovery Master Plan.* 1991.

Seoul Metropolitan City. *Seoul Plan for Mountain Scenic Landscape Conservation.* 2000.

Seoul Metropolitan City. *Seoul Plan for Mountain Viewing Landscape Conservation.* 2002.

Seoul Metropolitan City. *Seoul Basic Plan for Landscape Management.* 2005.

Seoul Metropolitan City. *Seoul Plan for Historic and Cultural Landscapes.* 2010.

Seoul Metropolitan City. *Basic Plan for Historic and Cultural City Management in Four Gates of Seoul.* 2012.

Shepherd, R. "The Baekdu Daegan as a Symbol of Korea." Accessed October 7, 2024. https://www.kocis.go.kr/eng/webzine/201810/sub02.html.

Song, I. H. *Maps of Old Seoul.* Seoul: Seoul Museum of History, 2016.

Statistics Korea. "Traditional Temple Designation Registration Status." Accessed April 28, 2024. https://www.index.go.kr/potal/stts/idxMain/selectPoSttsIdxMainPrint.do?idx_cd=2429&board_cd=INDX_001.

The Seoul Institute. *Look at Seoul through Maps 2000.* 2001.

Wybe, K. "The Nature of Urban Seoul: Potential Vegetation Derived from the Soil Map." *International Journal of Urban Sciences* 17, no. 3 (2013): 320–35.

Yoon, H. K. *P'Ungsu: A Study of Geomancy in Korea.* Albany: State University of New York Press, 2016.

이의성. "근대도시계획과정에서 나타난 공동묘지의 탄생과 소멸: Rise and Fall of the Seoul Public Cemeteries in the Modern Urban Planning Process." Master's thesis, Seoul National University, February 2021. https://s-space.snu.ac.kr/handle/10371/176329.

김기천. "[만물상] 시민아파트." *Chosun Ilbo*, November 29, 2005. https://www.chosun.com/site/data/html_dir/2005/11/28/2005112870404.html.

차현진. "현재에 충실하라." *Chosun Ilbo*, August 26, 2021. https://www.chosun.com/opinion/specialist_column/2021/04/08/2DPWH2UVPZBYHK3LFUYTVPRGUY/.

Imprint
판권

Seoul Mini-Mountains: Reading the City from Its Topographical Gaps
서울의 작은 산: 지형 격차로 도시 읽기

By
지은이
Géraldine Borio
제랄딘 보리오

Originally published in Paju, Korea, on June 27th, 2025
2025년 6월 27일 초판 발행, 대한민국 파주

Publisher
펴낸곳
Ahn Graphics;
Ahn Myrrh, An Mano, Oh Jinkyung (publisher), An Mano (director), Gu Minjeong (editor-in-chief), Lee Juhwa (editor), Park Hyunsun (designer),Kim Chaerin (marketer), Park Miyoung (manager)
안그라픽스;
안미르, 안마노, 오진경(펴낸이), 안마노(디렉터), 구민정(편집장), 이주화(편집자), 박현선(디자이너), 김채린(마케터), 박미영(매니저)

www.agbook.co.kr

Book Concept and Design
책 콘셉트 및 디자인
Typography Cabinet, Basel; Ludovic Balland and Annina Schepping
타이포그래피 캐비닛(바젤); 루도빅 발랑, 아니나 세핑

Texts
글
Géraldine Borio
제랄딘 보리오

Drawings
그림
Géraldine Borio with Tina Wang Tianduo, Kelly Mok Man Wing, Karl Lam, Laurene Cen, and (on pp. 160, 168) Keh Jessie Lok Yi, Lam Nga Yam Oddy, Leung Hiu Lam
제랄딘 보리오, 티나 왕 텐뭐, 켈리 목 만 윙, 칼 람, 로린 천, 그리고 (pp. 160, 168) 케 제시 록 이, 람 응아 얌 오디, 렁 히우 람

Photography
사진
Nils Clauss (pp. 11–38, 80, 88, 114, 120, 148, 150, 164, 166,168), Géraldine Borio (pp. 88, 90, 106, 114, 116, 122)
닐스 클라우스 (pp. 11–38, 80, 88, 114, 120, 148, 150, 164, 166, 168), 제랄딘 보리오 (pp. 88, 90, 106, 114, 116, 122)

Research Assistant
연구 보조
Tina Wang Tianduo, Kelly Mok Man Wing, Karl Lam, Laurene Cen
티나 왕 텐뭐, 켈리 목 만 윙, 칼 람, 로린 천

Copy Editor
카피 에디터
Cathy Coote
캐시 쿠트

Translator
옮긴이
Alice Do Ok-ran
앨리스 도옥란

Printing and Binding
인쇄 및 제작
Crein
크레인

Thanks to
Ludovic Balland, Annina Schepping, An Mano, Nils Clauss, Wang Tianduo Tina, Mok Man Wing Kelly, Karl Lam, Laurene Cen, Cathy Coote, Sanki Choe, Yiseo Yoon, Minjung Lee, Dongwoo Yim, Rafael Luna, Kangil Ji, Namjoo Kim, Seonju Kim, Susann Ahn, Myriem Alnet, Students from HKU Architecture who participated in the courses, Evelyne Gaümann, David Grünig and Jim.

ISBN
979.11.6823.100.9 (93980)